Interaction of Terahertz Radiation
With Semiconductor Lasers

Interaction of Terahertz Radiation With Semiconductor Lasers

RUHR UNIVERSITÄT BOCHUM
Germany

A PhD dissertation submitted to the Faculty of Electrical and Information Technology for the award of the doctorate degree in Electrical Engineering

Presented by

Msc. Jared Ombiro Gwaro

Birthplace(Kisii)

Prof.Dr.Martin R. Hofmann, Supervisor

Prof.Dr.Martin Koch, Co-Supervisor

May, 2019

Date of Examination: 11.06.2019

Bibliografische Information der Deutschen Nationalbibliothek
Die Deutsche Nationalbibliothek verzeichnet diese Publikation in der Deutschen
Nationalbibliografie; detaillierte bibliografische Daten sind im Internet über
http://dnb.d-nb.de abrufbar.
    1. Aufl. - Göttingen: Cuvillier, 2019
    Zugl.: Bochum, Univ., Diss., 2019

© CUVILLIER VERLAG, Göttingen 2019
   Nonnenstieg 8, 37075 Göttingen
   Telefon: 0551-54724-0
   Telefax: 0551-54724-21
   www.cuvillier.de

   ISBN 978-3-7369-7037-3
   eISBN 978-3-7369-6037-4

Abstract

Terahertz (THz) technology bears great potential in spectroscopy, imaging, material science, security screening and high-speed wireless communication. However, the generation of intensive, directional THz radiation has been difficult and the THz frequency range has long been considered the last final frontier of the electromagnetic spectrum. Recent advancement in optoelectronic terahertz generation techniques and high power electronic sources has helped to bridge the THz gap and has opened up a wealth of new applications for THz technology. However, there is still a major technical limitation in developing THz systems for mass markets, mainly due to the cost of THz hardware components including sources and detectors. In this regard, we investigated the use of semiconductor diode lasers as THz detectors as well as excitation sources for photomixers for THz generation. For THz detection, we investigated the interaction of semiconductor lasers with THz radiation. Intense THz radiation from different sources and at various frequencies was injected into the laser diode. The laser diode was operated in Littman configuration to ensure clean single mode operation in the near infrared. The charge carrier system in the semiconductor was expected to interact with the injected THz radiation and introduce nonlinear frequency mixing. This nonlinear mixing was to induce sidebands in the near infrared optical spectra and was to be analyzed with an optical spectrum analyzer. This may lead to the demonstration of a simple, cost effective and compact room temperature THz spectrometer since the distance between the emission line and the sidebands equals the incident THz frequency. Unfortunatly, due to unprecedented challenges the interaction of THz radiation with diode laser experiment was not successful.

Another approach was to demonstrate a compact and cost effective THz source based on monolithic distributed Bragg reflector diode laser emitting two frequencies simulteneously. We successfully demonstrated 300 GHz continuous wave THz radiation, with fiber coupled ion implanted photoconductive antennas used as photomixing devices. The generated THz radiation was tunable via temperature adjustments and current injection. This approach provided a coarse tuning in the range of 286 GHz to 320 GHz. We successfully demonstrated its potential use in non-destructive plant moisture measurements of a leaf induced to drought stresses and for moisture monitoring in drying process of pieces of paper. Due to the fact that the tuning of the developed THz source was coarse, we proposed the use of a new diode which was electrically tunable for fine tuning of the generated THz frequency. The new diode offers optical beat signal adjustments via carrier injection to the DBR section using micro resistor heater integrated on top of the DBR segment. The optical beat tuning via carrier injection was fast and offers tuning which should be free from mode hopping. The injection current to the resistor heater can be adjusted between 0 to 350 mA, an optical beat adjustment of between 100 GHz-300 GHz was realized. This bandwidth was only limited with the overlap of the two modes at higher heater currents of 250 mA to 350 mA. THz radiation emission via photomixing in the range 100 GHz-300 GHz was successfully demonstrated, these results were in good agreement with the optical beat signal measurements. Finally, a simple spectrometer suitable for THz metrology measurements such as thickness determination of Polyethylene sample (PE) was realized and also its application in THz spectroscopy was demonstrated by the determination of the spectroscopic transmission characteristics of a THz filter. In summary, two compact THz sources emitting at 300 GHz were successfully demonstrated and this was a major milestone towards development of compact and cost effective THz system for mass market application.

Zusammenfassung

Die Terahertz-Technologie (THz) bietet ein großes Potenzial in den Bereichen Spektroskopie, Bildgebung,Materialwissenschaft, Sicherheitsprüfung und drahtlose Hoch geschwindigkeitskommunikation. Die Erzeugung von intensiver, gerichteter THz-Strahlung war jedoch schwierig und der THz-Frequenzbereich wurde lange als das Ende des elektromagnetischen Spektrums angesehen. Jüngste Fortschritte in der optoelektronischen Technologie hat eine neue Anwendung für die THz-Technologie eröffnet. Bei der Entwicklung von THz-Systemen für Massenmärkte gibt es jedoch immer noch eine große technische Einschränkung, hauptsächlich aufgrund der Kosten für THz-Hardwarekomponenten einschließlich Quellen und Detektoren. In diesem Zusammenhang betrachten wir die Verwendung von Halbleiter-Diodenlasern als THz-Detektoren sowie Anregungsquellen für Photomischer zur THz-Erzeugung. Für die THz-Detektion haben wir die Wechselwirkung von Halbleiterlasern mit THz-Strahlung untersucht. In die Laserdiode wird intensive THz-Strahlung von verschiedenen Quellen und bei verschiedenen Frequenzen injiziert. Die Laserdiode wurde in Littman konfiguration betrieben. Durch THz-Strahlung und nichtlineare Mischprozesse sollen, Seitenbänder im den optischen Spektrum im nahen Infrarot induziert und mit einem optischen Spektrumanalysator analysiert werden. Dies wave ein effizientes THz-Spektrometer, da der Abstand zwischen der Emissionslinie und den Seitenbändern der einfallenden THz-Frequenz entspricht. Leider war das Experiment aufgrund zu schwader weschselring von THz-Strahlung mit Diodenlasern nicht erfolgreich.

Ein weiterer Ansatz bestand darin, eine kompakte und kostengünstige THz-Quelle zu demonstrieren, die auf einem monolithischen Distributed Bragg-Reflektor-Diodenlaser basiert, der zwei Frequenzen gleichzeitig emittiert. Wir konnten erfolgreich 300-GHz-Strahlung demonstrieren, wobei als photomischende Vorrichtung fasergekoppelte ionenimplantierte photoleitende Antennen verwendet wurden. Die erzeugte THz-Strahlung war über Temperatureinstellungen und Strominjektion einstellbar. Dieser Ansatz ermöglichte eine grobe Abstimmung im Bereich von 286 GHz bis 320 GHz. Wir haben die mögliche Verwendung des entwickelten Systems bei zerstörungsfreien Anlagenfeuchtigkeitsmessungen an einem Blatt, das durch Trockenheit ausgesetzt wird, und zur Feuchtigkeitsüberwachung beim Trocknen von Papierstücken erfolgreich demonstriert. Aufgrund der Tatsache, dass die Abstimmung der entwickelten Quelle grob war, haben wir die elektrisch zur Feinabstimmung der erzeugten THz-Frequenz einstellbar ist. Die neue Diode bietet optische Schwebungssignalanpassungen über die Trägerinjektion in den DBR-Bereich mithilfe eines auf dem DBR-Segment integrierten Mikrowiderstandheizers. Das Tuning über ladungsträger-Injection ist schnell. Der Injektionsstrom in der Widerstandsheizung kann zwischen 0 und 350 mA eingestellt werden, es wurden Differenz frequenzen später eine Diode verwendet zwischen 100 GHz und 300 GHz realisiert. Diese Bandbreite war nur durch die Überlappung der beiden Modi bei höheren Heizerströmen von 250 mA bis 350 mA begrenzt. Die Emission von THz-Strahlung durch Photomischung im Bereich von 100 GHz bis 300 GHz wurde erfolgreich demonstriert. Diese Ergebnisse stimmten gut mit den Messungen des optischen Schwebungssignals überein. Schließlich wurde ein einfaches, für THz-Messungen geeignetes Spektrometer für die Bestimmung der Dicke einer polyethylene-Probe (PE) realisiert, und auch seine Anwendung in der THz-Spektroskopie wurde durch die Bestimmung der spektroskopischen Transmissionseigenschaften eines THz-Filters demonstriert. Zusammenfassend wurden zwei kompakte THz-Quellen, die bei 300 GHz emittierten, erfolgreich demonstriert. Dies war ein wichtiger Meilenstein auf dem Weg zur Entwicklung eines kompakten und kostengünstigen THz-Systems für den Massenmarkt.

Dedication

This dissertation is dedicated to my loving mum and my late Dad. For the sacrifices they made to raise me up and train me to be whom I am today. Your love, prayers, encouragement and motivation have helped me to come this far.

Acknowledgements

I would like to acknowledge several people who made their contribution in one way or the other towards supporting me to finalize this dissertation and due to their unwavering support I managed to endure the difficult times associated with a PhD study.

I'd like to sincerely thank Prof.Dr. Martin R.Hofmann for giving me an opportunity to join his research group and offered me invaluable support to learn new ideas in a challenging new field of Terahertz technoloy over the past four years. Your support can not be paid by any means it is invaluable. It has been a pleasure to work with you and I have learned a lot by doing this PhD under your supervision.

My colleagues at the Photonic and Terahertz research group for providing me with a good working environment. Many thanks to Dr.-Ing. Carsten Brenner for the day to day consultation whenever I had technical challenge through out this research work. Your support was enormous and I say thank you, for you were willing to assist whenever I encountered a technical problem and you were ready to help. Dr.-Ing. Nils C. Gerhard for invaluable comments during the PhD colloquium which lead to the rethinking on areas to improve and eventually lead to shaping up of my research approach. Many thanks to Nils Surkamp for the design of the base plate and submount for the compact external cavity semiconductor diode for the detection experiment and fruitful discussion during the design of the Y-shaped laser diode submount.

I would like to thank Prof.Dr.Martin Koch for accepting to lend us the Gunn diode for the interaction experiment and made arrangement which enabled us to access THz-TDS system in Marburg for further experiments. Am also greatful for the team in Marburg lead by Dr.Sebastian Gies and Phillip Richter for accepting to work with us when we visited Marburg for the interaction experiments with the use of the THz-TDS system. Thanks too to Dr. Ksenia for showing me the operation of Gunn diode.

I like also to sincerely thank my parents my late Dad and mum for sacrificing so much to educate me and the support you accorded me throughout my career life with whethever you could afford. Thank you very much for your support. I have no better words to say thank you. I would also like to thank my wife and my daugthers Miriam, Lisa and Claudia who endured hard times while I was busy writting this thesis and had little time for them.

I would like to acknowledge financial support from the kenyan government and DAAD for the joint kenya-DAAD scholarship. Further a lot of thanks to Maasai Mara University for granting me an opportunity to go on study leave.

Lastly, I thank you all my friends I have not mentioned for your unwavering support in one way or other.

Finally I thank God for the gift of life and good healthy during this studies.

Contents

List of Figures

List of Tables

Chapter 1

Introduction

The terahertz (THz) spectral region (0.1-10 THz) remained unexplored part of the electromagnetic frequencies for quite sometime hence was commonly described as the forgotten gap or THz gap. This was mainly attributed to scarcity of excellent THz emitters and receivers in the THz spectral region. However, in the recent years, a lot of research attention has been witnessed in this spectral region leading to great research progress devoted to development of novel THz sources and detectors [1]. The advance in optoelectronic THz generation and detection techniques has led to an increased interest for THz applications in many existing and emerging fields such as medicine, security, manufacturing, basic sciences, and communications. Such great attention in the THz radiation and its potential applications has taken the THz research a notch higher with a lot of activities devoted towards development of cost effective and efficient THz emitters as well as THz detectors with high sensitivity [2].

Considerable approaches to detect THz radiation have been developed. This includes, detectors such as Schottky diode, photoconductive antenna and electro-optic detection and thermal detectors [3], [4]. Pyroelectric detectors, Bolometers and Golay cells are the conventional THz detectors, which belong to the class of thermal detectors and provides broadband spectral response, however they are large in size, and have slow sensitivity due to thermal detection mechanism of heating a material before detection is achieved and can only measure the intensity of THz radiation [4], [5]. In contrast, the photoconductive or electro-optic detection can provide direct access to temporal THz transients and thus provide both phase and amplitude of a THz wave which is desired information for high spectral resolution spectrocopy and imaging applications. But usually, this detection concept requires rather complex measurement schemes. So, in summary, the realization of efficient THz sources and detectors is still probably the most challenging part of THz technology.

In general, THz radiation can be categorised into two groups namely: pulsed THz source commonly referred as broadband and continuous wave (CW) THz sources also known as narrowband sources. Different mechanisms are involved in the generation of pulsed THz sources as well as CW THz sources. For pulsed systems for example, by exciting a semiconductor material with ultra-short laser pulses one can generate THz transient, by ultilizing nonlinear mixing processes in non-linear crystals or photoconductive antennas. The generation of THz transient with ultra-short pulse can provide a THz radiation with broad spectrum of THz frequencies. CW THz sources on the other hand can be generated with differnt means such as electric source, Gunn diodes, THz gas lasers, photomixers, difference frequency generation, quantum cascade lasers as well as free-electron lasers. CW THz emitters generates a THz wave signal with a narrower linewidth when compared to pulsed THz sources however with a higher spectral THz power. Unlike in pulsed system where power is distributed throughout the entire spectrum, in CW THz power is concentrated on a single frequency. CW THz sources are highly desired in high-resolution THz spectroscopy, imaging as well as in wireless communications [2].

THz spectroscopy requires usually THz radiation with wide frequency range which is tunable and has sufficient power levels in the range of microwatts. In addition, THz source having small size, low cost and room temperature operation condition will be highly suited for such application [3]. Some of the systems mentioned above can generate relatively high emission power. However, they are large in size, cost a fortune and their tunable frequency range is limited [6]. It is still technologically challenging to develop compact CW THz systems that can provide broad tunability of frequencies throughout THz spectral range with sufficient power levels. Furthermore, development of table top THz sources would be a major milestone towards enabling THz technologies to be transferred from the lab setting into real market applications [2].

In order to realize such extremely efficient, inexpensive and compact CW THz sources, photomixing of two wavelengths based on diode lasers is currently one of the promising potential techniques. Moreover, it has been demonstrated in the past that it is possible to generate THz radiation from a two color diode lasers without external mixing elements directly at ambient temperature conditions [7], [8]. This thereby provides an extremely compact THz source and cheap system operating at room temperature but the THz power generated with such a system was in a few magnitude. Hoffmann et al., analyzed the restricting factors to access the THz bandwidth as well as the semiconductor nonlinearity of a two color diode in generating THz signal. The results demonstrated that the gain medium doesn't limit the bandwidth, since the operation of the diode was possible up to 30.2 THz optical beat frequency and even the carrier response exhibited a nonlinear response of up to 4.2 THz [9]. Further experiments and theoretical analysis have suggested some technological efforts to enhance the power generated from a two color diode lasers to application levels by improving on the design of the diode structure [8]. However, the confirmed nonlinear response of the carrier system to THz frequencies introduces the possibility for further applications, such as THz detection using laser diodes. This would be of high interest in the development of a whole THz system, which demands that both a THz emitter and a THz receiver to be as small as possible. Recently, it was indeed demonstrated in first proof of principle studies [8] [10], that semiconductor diode lasers can not only generate but also, they can detect THz radiation. The realization of diode laser based THz sources and detectors, will offer the prospects of building a very compact and complete THz systems on a single optoelectronics substrate [8].

The motivation in this dissertation was to investigate how THz radiation interacts with the charge carriers in diode lasers and, in particular, to demonstrate THz detection using semiconductor lasers. The experimental work was to realize an optimized detection scheme enabling THz spectroscopy using commercially available semiconductor lasers diode. This could add new potential capabilities applications in fields such as security sensing, spectroscopy, imaging, material research and communications. Further work, was to implement compact and cost efficient THz sources based on two color semiconductor diode lasers via the process of photomixing and contribute in development of THz systems with capability of enabling the transfer of THz technology from the lab setting into real world application.

1.1   Organization of this dissertation

This dissertaion is structured into six chapters. The general introduction of this dissertation and the motivation of doing this research investigation is provided in chapter 1. In chapter 2, we describe the fundamentals of the semiconductor diode as well discuss the operation of semiconductor diode laser. A few types of diode lasers more so the ones used in this study are highlighed with their key features and operation principles discussed. In chapter 3, different detection schemes and how they operate with an analysis on the need to develop new compact detection approaches in order to make THz detection available for application is given. Chapter 4, describes the study of the interaction of THz radiaton from different THz sources with a semiconductor diode arranged in Littman configuration. The motivation being to demonstrate a compact detection scheme of THz radiation with diode laser and preliminary findings on our experiment are given. In chapter 5, the fundamentals and mechanism of THz generation with photoconductive antennas is discussed. The implementation of compact THz sources based on the use of monolithically integrated DBR lasers emitting two modes simultaneously as an excitation source for photoconductive antenna is discussed. In this study two laser diodes are discussed, One was tunable via temperature changes of the laser chip plus the diode submount while the second was electrically tunable via carrier injection to the laser chip alone. More details on the operation of these devices will be given and their potentail use for CW THz generation is presented. Finally the results on the developed THz systems will be given with potential application in non-destructive moisture measurements, moisture monitoring on drying process of pieces of paper, THz spectrocopy such as characterization of THz filter and THz metrology such as thickness measurement of polyethylene sample will be discussed. Lastly, a concise summary of the entire dissertation is described in chapter 6.

Chapter 2

Fundamentals

In this chapter, semiconductor diode laser which is one of the key components used in the study carried out in this dissertation is discussed. The fundamental aspects and operation of a semiconductor diode laser are presented. A few types of diode lasers, more so ones used in this research are introduced with their operation principle and key features highlighted.

2.1 Semiconductor diode laser

A semiconductor laser is an electronic device that radiates light signal when a forward bias is applied across its p-n junction resulting to the flow of a current. Semiconductor lasers were first developed in the earlier 1960's [11] and over time diode lasers have developed into key components in modern optoelectronics and photonics technology. In our day to day applications, diode lasers have found their way into compact disc systems (for data storage), pointers, printers, bar code scanners, and optical communications for the transmission of information through optical fibers. Furthermore, they have gained use in the studies of atomic structure and quantum-mechanical effects and more recently they are gaining applications in other emerging fields such as THz generation and detection [8] [10] [12],Raman spectroscopy [13], photoacoustic imaging [14], optical coherence tomography [15] and Biophotonics [16]. Our focus here was to use semiconductor diode lasers as key development tools to generate THz signal as well as a THz detector. The following features distinguishes diode lasers from other lasers: They are compact, diode lasers are quite small in size, the small size footprint enables them to be easily integrated into many instruments. They are highly efficient (around 50% efficiency) which makes them to be driven by low injection currents. As they can be manufactured in large quantities they are cheap. Furthermore, mass production of diode lasers is cost effective hence they are affordable for various applications. Finally, semiconductor diode lasers can be tailored to emit particular wavelengths. Most lasers are usually limited to a specific atomic transition, therefore their emission wavelength is defined. While, diode lasers can be designed to emit wavelengths covering the broad frequencies of the electromagnetic spectrum [16], depending on the composition of the different combinations of the elements of group III-V semiconductors. Table 2.1 gives some common alloys on semiconductor substrate and their corresponding laser wavelength emission.

Compound	Spectral range	application
$Al_xGa_{1-x}N$	uv	
GaN	uv 350 nm	
$In_xGa_{1-x}N$	blue 400-480 nm	data storage, display
$In_xAl_yGa_{1-x-y}P$	red 615-750 nm Ref [15]	display
$Ga_xIn_{1-x}P(x = 0.5)$	670 nm	display
$Ga_xAl_{1-x}As(x = 0 - 0.45)$	620-895 nm	
$Al_xGa_{1-x}As_{1-y}P_y$	670-890 nm Ref [15]	
GaAsP	785 nm	Raman spectroscopy Ref [13]
GaAs	904 nm	diode pumping
$In_xGa_{1-x}As(x = 0.2)$	980 nm	solid-state and fiber lasers
$In_xGa_{1-x}As_yP_{1-y}$	1110-1650 nm	
(x=0.73, y=0.58)	1310nm	communication
(x=0.58, y=0.9)	1550nm	

Table 2.1: Table of semiconductor compounds for the manufacture of different diode with different wavelength emission [17].

The selection of different or a particular wavelength emission for a certain compound structure is perfomed by course adjustment of the various combinations of the molfunction x, y and z. With several combinations of group III-V compounds a broad spectral range can be covered [15]. In this study we employ structures made of GaAsP/GaAs emitting at 785 nm (for THz generation experiments) and GaAs/AlGaAs emitting at 840 nm (for THz detection experiment) respectively.

2.1.1 Energy bands

In a semiconductor material, the interactions of atoms split the discrete energy levels into distinct energy levels, which are closely spaced in energy forming a continuous band of energy state commonly referred as energy band. The electrons can occupy this energy bands. In thermal equilibrium, the uppermost band which is empty of electrons is the conduction band (CB). The CB is distinct from the next lower level called valence band (VB) which is full of electrons, by a transition energy referred as bandgap energy E_g. Electrons can move from the VB to CB, when a semiconductor materials is supplied with an energy exceeding the E_g [18].

At absolute zero, Fermi levels are used to describe the energy occupance of electrons. A semiconductor material with majority of holes is called p-type and has a Fermi level E_{FP} which is close to the VB while a semiconductor material with majority of electrons is an n-type and its Fermi level E_{FN} is close to CB. The joining of p-type semiconductor material back to back with an n-type semiconductor materials forms a sandwich commonly referred as a p-n junction. A p-n junction is what is termed a diode. Figure 2.1 summarizes the basic operation of a diode depended on the applied bias. At zero bias Figure 2.1(a), the Fermi level across the diode is continuous therefore the Fermi levels are equal $E_{FP} = E_{FN}$. The free electrons and holes in the junction diffuse resulting to a zone free of electrons and holes forming the so called depletion layer. An in build voltage V_o which is generated as a result of this recombination prevents the electrons and holes from transiting from one side of the junction to the other. Across the junction if a forward bias V

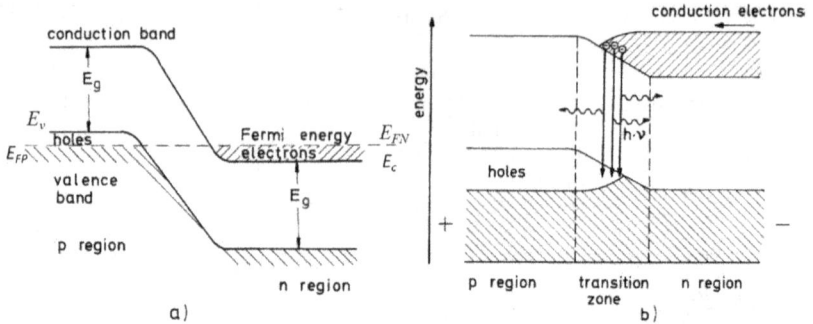

Figure 2.1: The energy band diagram of a p-n junction (a) when not biased and (b) when forward biased [21].

is applied, the Fermi level is split into Quasi-Fermi levels, and is not continuous any more [19] [20]. Therefore the electrons and holes are free to transit the depletion region. As electrons recombine with the holes, they release a photon of energy $h.v$ when stimulated emission occurs as shown in Figure 2.1(b) [21]. Over time a p-n junction was found not to be efficient enough hence resulted to the development of another structure the so called double-heterostructure [23]. This structure is based on an active laser material sandwitched between a material with large bandgap. As a result the charge carriers and the photons are constricted into the active region which makes the laser diode to be operated with low injection current [24].

One of the most widely developed diode lasers are based on the GaAs/AlGaAs semiconductor material. They comprise of a GaAs active layer embedded within layers of AlGaAs which serve to confine the laser radiation. The two opposite faces are then smoothened, cleaved, and made parallel with respect to each other. These facets, act as resonant mirrors which provide the feedback desired for laser action. For laser action to occur the p-n junction is pumped by an electrical current commonly referred as the injection current. At the center of the diode structure is the active region, when an injection current passes other layers and reaches this region, electrons are accelerated and transit from the VB to the CB, in the process a population inversion is created resulting to emission of photons when the electrons and holes under go stimulated emission [22]. The burying layer or current blocking layer restricts the current to a narrow strip which ensures low current injection for laser action with a single lateral mode [20]. A schematic structure of semiconductor laser with a double heterostructure is shown in Figure 2.2.

The electrons distribution in the conduction band and valence band is governed by their Quasi-Fermi energy which is determined by their population. The probability of electrons occupying a particular energy levels in the conduction band is described by the Fermi-Dirac function [2, 24]:

$$f_{cb}(E) = \frac{1}{1 + \exp\left(\frac{E - F_n}{k_B T}\right)} \tag{2.1}$$

and in the valence band

$$f_{vb}(E) = \frac{1}{1 + \exp\left(\frac{E - F_p}{k_B T}\right)} \tag{2.2}$$

where E_v and E_c are the allowed energy states in the VB and the CB respectively, and F_n and F_p are their corresponding Quasi-Fermi energy levels, T is the absolute temperature and k_B is Boltzmann constant.

A particular Quasi-Fermi level position depends on the temperature and the density of electrons and holes. Should the following condition

$$E_g < E_2 - E_1 = \frac{h}{2\pi}\omega < F_c - F_v \tag{2.3}$$

be satisfied, the recombination of charge carriers will dominates the absorption and other losses hence laser action will take place due to stimulated emission [2]. Another important requirement to be fulfilled for stimulated emission to be observed is the active region must be a direct band gap seimconductor where stimulated emission can occur. In addition to that, the charge carrier population within the active region should be sufficient such that their Quasi-Fermi levels are discrete by an energy greater than that of the emitted radiation [11].

The semiconductor band structures are categorized into two groups. The first band structure is a direct bandgap structure, which is the band in which the maximum energy level of the valence band falls within the same wavevector k space or at same crystal momentum with the minimum energy state of the conduction band. The transition of electrons takes place along the zero point of the wavevector ($k = 0$) space with the momentum conservation law satisfied. The second band is indirect bandgap structure in which the minimum energy level of the conduction

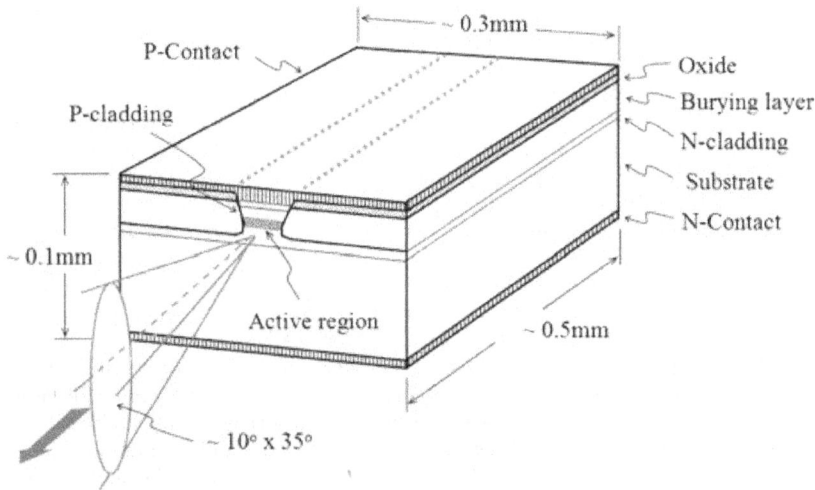

Figure 2.2: Schematic of typical laser diode chip, cladding is AlGaAs, active region is GaAs. Burying layer restricts current flow to narrow stripe perpendicular to the slit, through active region [25].

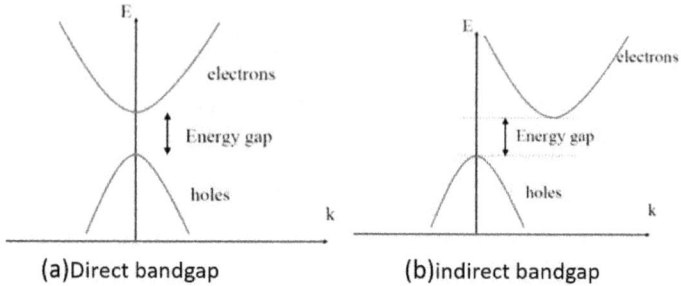

Figure 2.3: The bandgap materials types in semiconductor structure [26].

band doesn't lie in the same vector space with highest energy state of the valence band. This makes it challenging to obtain a transition of electrons easily hence even with alot of current pumping it is impossible to achieve stimulated emission in indirect band gap. Most of group III–V compounds such as InGaAsP, AlGaAs, InP,and GaAs belong to the direct band gap material, while Si and Ge are materials in the indirect category [17, 27]. Figure 2.3 shows the schematics of the two band gap transitions in wavevector space.

2.1.2 Different semiconductor laser structures

Semiconductor lasers (SLs), can be directly pumped and emit photon either from the edge or from the surface of their structure. The structures in which the emission of the light signal is from the top surface with the direction of the signal being perpendicular with respect to the surface substrate are said to be surface emitting lasers examples of lasers in this category are the vertical cavity surface emitting lasers (VCSELS). While structures that enables the light signal to travel along the wafer surface of the semiconductor substrate are the edge emitting lasers. Due to the differences in refractive index between the burying and the active layers, the carriers and the photon are confined along the active layer region [15]. With gain guided structures(i.e with restricted current pumping) or by implementing index-guided structure in which lateral step index is designed one can obtain lateral confinement of the optical modes. However, with a lot of progress in growth techniques, quantum well (QW) structure with reduced active layer thickness have been realized. They exhibit low threshold current and excellent modulation features. Most of the available diode lasers in the market are made from QW structures [2].

The structures can emit either a single mode or multiple modes depending on the number of wavelengths they are able to select. The Fabry Perot Laser (FP-laser) structures have no wavelength selection devices hence they usually emit multiple wavelengths. This causes the laser to lase at higher modes and permits the laser to oscillate between different longtudinal modes. In such structures the laser exhibits low coherence and large linewidths. In order to avoid this, it is necessary to have wavelength selection in the laser under which it operates. This can be achieved by inserting a wavelength selective optics to select only single frequency and eliminate other frequencies. This is achieved through incorporating periodic gratings struc-

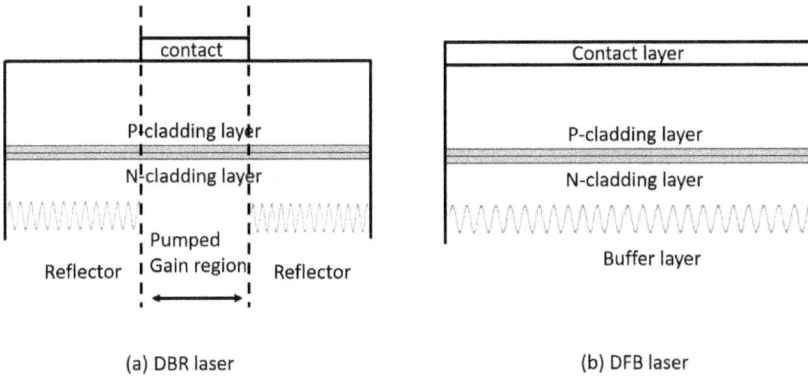

Figure 2.4: Schematic diagram of (a) DBR laser and (b) DFB laser semiconductor structures [24].

tures within the laser waveguide which enables a single laser emission at any given time. Such structures are grouped into; Distributed Feedback (DFB) lasers which are structures which incorporate grating structures in the active region, and Distributed Bragg Reflector (DBR)lasers which are structures which incorporates the grating in the passive region [19, 24]. DFB and DBR lasers operate in a single frequency even under pulsed mode with narrow linewidth and highly coherent signal, in contrast to FP-lasers, which exhibit several wavelengths when operating in pulsed mode [19]. We employed two DBR lasers structures emitting two colors simulteneously at 785 nm center wavelength to demonstrate THz generation. Figure 2.5 shows a vertical layer structure of the two-color DBR laser used in this study. The active region was made up of GaAsP semiconductor material which was sandwitched between two waveguide layers made of AlGaAs and two cladding layers. The AlGaAs layer has a thickness of 500 nm while the cladding layer is $1\mu m$. The whole structure is grown on GaAs substrate by metal organic vapour phase epitaxy (MOVPE). More details on the design and fabrication of this structure are described in ref [28, 29]. We employed such structures to demonstrate THz emission via photomixing in ion implanted log-spiral photoconductive antenna. Detailed investigation of the use of this structures for applications in THz generation is discussed in chapter 5.

DFB and DBR structures can be tunable with either temperature (course tuning) or via current (fine tuning) which enables the refractive index to be varied along the active region in the process changing the emission wavelength of the laser. Wavelength selection via temperature may result to instabilites due to mode hopping and usually tunability of such structures is limited. Wide tuning of the emission frequency in the SLs can be achieved not with DFB/DBR but with FP-lasers by employing external cavity configuration, incorporating the use of diode laser as the laser gain medium, wavelength filter for selection of the laser mode and an end mirror to provide the optical feedback. There are two main categories of external cavity configurations namely the Littman-Meltcalf configuration and the Littrow configuration. A diffraction grating is commonly used as the wavelength selector and enables a broad tuning across the gain region of the diode. A semiconductor

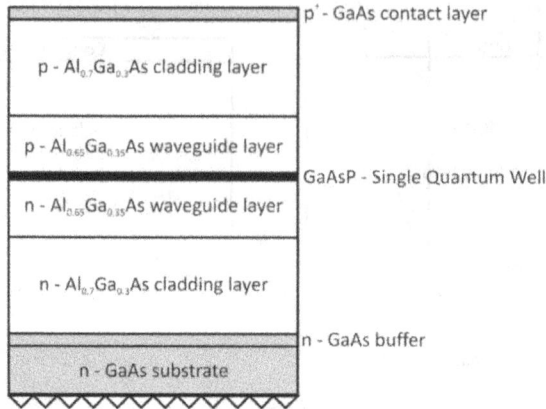

Figure 2.5: Schematic illustration of vertical layer structure of two-color DBR lasers used in this study [28].

laser having its front facet deposited with an antireflection coating (AR) while its rear facet is deposited with high reflecting (HR) coating is used as the gain medium. An optical feedback from the diffraction grating is sent back into the diode through the lens and consists of reflected mode with narrow linewidth than the beam which was incident at the grating. Stimulated emission is induced by the reflected signal in the diode gain medium at the wavelength of the feedback beam. The diode then lases at the wavelength of the feedback signal at a much narrower linewidth better than that of the incident signal [30]. Figure 2.6 shows the two external cavity configurations. In Littrow configuration, mode selection is done by the adjustment of the grating while in Littman-Meltcalf configuration the wavelengths are selected by tuning the additional end mirror introduced in the cavity while the diffration grating remains fixed [31].

As mentioned before, the grating acts as a wavelength selector providing the desired optical feedback. Figure 2.7 shows a blazed grating in which a light signal is incident at the diffraction grating. The angle of incident is α, while the angle at which the beam is diffracted is β. In Littrow configuration, the angle of incidence equal to that of reflection. Therefore at any given time the order of the diffrated beam is sent back and travels along the path of the incident beam. According to the grating equation the following condition should be met for constructive interference to occur [30].

$$d(sin\alpha + sin\beta) = m\lambda \qquad (2.4)$$

where d is the spacing of the grooves, λ is the wavelength of the incident signal and m is the order of diffraction of the beam. In Littrow configuration $\alpha = \beta$, therefore equation (2.4) reduces to

$$m\lambda = 2dsin\alpha \qquad (2.5)$$

The blaze wavelength λ_B, which is the wavelength where much of the optical power or energy is concentrated which is commonly referred as the first order diffracted wavelength. The blaze angle θ_B is equal to β in Littrow configuration and hence the first order blaze wavelength is given by

$$\lambda_B = 2dsin\theta_B \tag{2.6}$$

Both configurations have advantages and disadvantages, Littrow configuration designs results to high emission powers in comparison to Littman design. However due to the tuning with the diffration grating in Littrow design, it is challenging to get a fixed output position all the time of tuning. Hence it is an advantage to use end mirror in tuning in Littman design as the output path of the output beams remains unchanged [30, 31]. In the investigation for use of semiconductor diode laser as a THz detector, we employed a Littman configuration, more details on this design will be discussed in chapter 3.

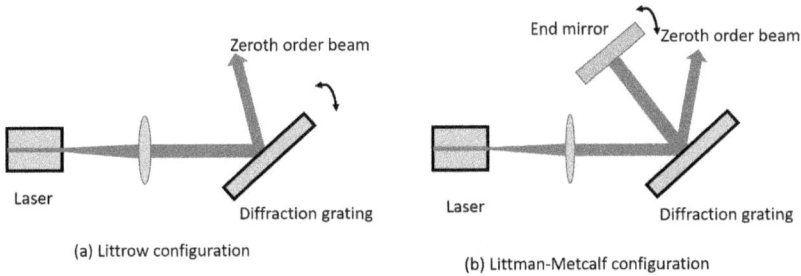

Figure 2.6: Schematic illustration of (a) Littrow configuration and (b) Littman-Metcalf configuration.

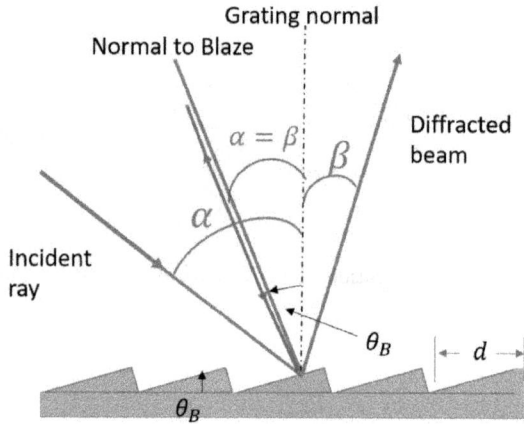

Figure 2.7: Schematic illustration of diffraction of beam in general case (red ray) and Littrow configuration blue ray in a blazed grating [30].

Chapter 3

THz detection

In this chapter, we discuss different approaches of THz detection and give an analysis for the need to develop new compact detection approach inorder to make THz technology available into many applications.

3.1 Introduction

The development of highly sensitive, compact, and affordable THz detection schemes operating at room temperature is crucial for realizing the wide spread of THz technology. However, THz region(0.1-10 THz) has remained elusive to conventional scientific approaches used in well-established neighboring regions of microwave and optical frequencies [32]. The challenge in THz region is attributed to comparatively low photon energy of THz radiation which is 4.1 meV at 1 THz. This is in order of 100 times less than that of visible light hence making it stubbornly difficult to detect [3]. Hence, it is quite complex and hard to develop THz detectors in the terahertz gap.

Nevertheless, the development of different detection schemes in the past, ranging from direct thermal detectors (Bolometers, Golay cells, Pyroelectric detectors) and indirect ones such as electro optic sampling and photoconducting antennas have contributed greatly to opening up THz region to scientific applications [33]. Bolometers, Golay cells, Pyroelectric detectors, are widely used thermal detectors and their major drawback arises from thermal processes involved in the detection which are not selective as they can detect even the surrounding heat originating from nearby instruments [34]. Furthermore, their sensitivity worsens in a hot environment, therefore the need for cryogenic cooling. In the absence of cryogenic cooling, the detector introduces electronic noise which when becomes dominant could make the detector less sensitive [35]. Bolometers are most sensitive thermal detectors, as they operate under cryogenic condition to Helium liquid temperatures. However, cryogenical cooling these detectors is expensive and results to their large size. Golay cells and Pyroelectric detectors, exhibit slow response time and low sensitivities despite them operating at room temperature [4]. Electro-optic sampling and photoconductive detection require complex set up and are difficult to implement.

3.2 Classification of THz detection schemes

THz detection methods are classified into two main categories namely: coherent and incoherent detection techniques. If the detector measures only the intensity, then it is said to be incoherent or direct detector. While, if the detector measures phase of the THz radiation as well as its amplitude then it is said to be coherent detection. Coherent THz detection is analogous to THz generation process, in that the same components and beam used for THz generation are the same used for detection. In coherent detection a THz signal is mixed with the optical signal used for THz generation and both are then detected, before the THz signal is extracted from the obtained measurement [36, 37]. Various methods for THz detection are discussed in the proceeding sections.

3.2.1 THz thermal detectors

Thermal detectors are a class of incoherent detection which covers a broad THz range and allows the detection of the intensity only. They operate on a simple mechanism, that a material absorbs incident radiation resulting to it being heated up. Their corresponding temperature increase produces some physical change such as resistance, pyroelectric voltage or thermoelectric voltage which can be measured by any temperature-dependent means [38]. Thermal detectors are usually manufactured from a THz absorbing material which is connected with a thermometer, with both the absorbing material and the detecting mechanism designed of the same material. The detector in this case is a thermometer normally suspended on lag and connected to a heat dissipating material [32, 39]. Figure 3.1 shows the general sketch of a thermal detector. Thermal detectors are generally the same the only chief difference between them is the technique employed to determine the temperature of the absorbing material. In pyroelectric detectors, the polarization change is measured as an electric current, in thermopile detection the variation of thermoelectric effect is measured, while in Golay cell, the detection is accomplished with the change on dilation of gas, whereas for bolometers an electrical resistance change is obtained [39]. The most predominat thermal detectors are bolometers and Golay cells. A bolometer operates under cryogenic condition and usually is cooled with liquid Helium. This enhances their sensitivity thereby detecting even very low THz signals. Golay cells and pyroelectric detectors both function well at room temperature condition. Bolometers and Golay cells, both can detect a broadband range of spectral frequencies and due to the fact that the detecting element material as to be heated up for a response to be noticed, they exhibit slow response time [5]. In this thesis, we introduce briefly the commonly used thermal detectors in the field of THz research, namely, pyroelectric, bolometer and Golay cells.

Pyroelectric detector

In Pyroelectric detector, the dielectric properties of a material placed in a small voltage biased capacitor, are changed when a THz radiation is incident on the material. The temperature change results to the appearance of charges in the capacitor which induces the generation of a current which is determined. The pyroelectrical coefficient properties of the detecting element and its temperature changes determines the amount of the current generated. The THz beam incident at the detector

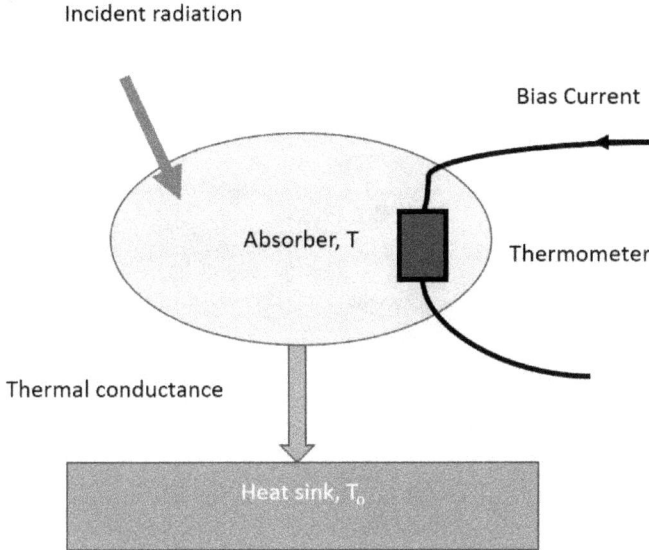

Figure 3.1: Scheme illustrating THz detection with thermal dectector.

can be modulated in order to improve the sensitivity of the detector. The detector suffers from amplifier noise which limits its sensitivity. The designing of a small detection area and its response speed engineered for fast laser pulse detection can be done to achieve fast thermal response time with this detector. This however, can result to the decline in the sensitivity of the detector due to reduced area of radiation interactons [38].

Bolometer

Bolometers are class of thermal detectors which operate at very low temperatures as mentioned before hence they are very sensitive to any small incident of THz radiation. The bolometer measurements are based on a thin resistive element which is designed from a material having low heat capacity and large absorptivity. The oxides of cobalt, manganese and nickel are some of the examples of materials used to build thermistor type bolometers. The resistive absorber material measures the change in temperature when a THz radiation is incident on it. Bolometers are broadband detection systems which can cover extremely wide range of the electromagnetic spectrum. They have high sensivity as they operate under cryogenic condition. However, they too suffer from slow response time since the heating and cooling of the detecting element are slow processes [32].

Golay cell

Golay cells detection mechanism is based on the gas volume changes in a chamber which results due to the incident radiation. The essential features of the Golay cell

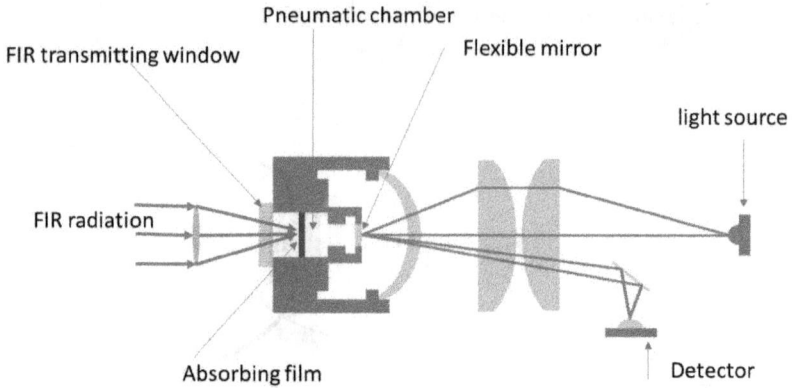

Figure 3.2: Scheme of THz detection with Golay cell [37].

include pneumatic chamber filled with a gas having low thermal conductivity such as xenon, a thin absorbing film, a light source and a photodiode. The detector is arranged such that the modulated incident radiation on the absorbing film heats up the gas in the pneumatic chamber, so that expansion of the gas distorts a flexible mirror. The mirror movement as a result of the gas expansion, deflects a light signal shining on a photodiode and therefore, produces a photodiode current change as the output. Figure 3.2 shows the main mechanism features for THz detection with Golay cell. The noise resulting from the exchange of thermal energy from the absorbing film to the gas in the chamber is the only factor which can limit the performance of the detector, usually the detector is highly sensitive with detectivity in the range $D = 3 X 10^9 cmHz^{1/2}W^{-1}$ can be obtained. They however have a long response time due to the thermal detection processes involved. [32, 37].

3.3 Coherent detection

3.3.1 Heterodyne and homodyne detection

In heterodyne detection, a THz signal is down-converted into a new signal commonly referred as intermediate frequency (IF) when it is mixed with a similar frequency of a local oscillator. Nonlinear elements like a Schottky barrier diodes are usually used as mixers in heterodyne detection. They mix the THz signal V_{THz} and the local oscillator signal V_{LO} generating a new signal at the intermediate frequency $V_{IF}=(V_{THz}-V_{LO})$ [32]. In comparison to thermal detection this detection approach provides a good sensitivity at room temperature operation. However, it requires an impedance matching to be able to achieve high Noise equivalent power (NEP) values and the bandwidth attainable is depended on the mixer signal that is IF and NEP is limited due to bandwidth of IF [39]. Heterodyne detection is the preferred approach in field which require high spectral resolution such as astronomy [8]. For homodyne detection, same components and optical beat signal employed to generate THz signal are used for detection. For detection if for example a photoconductive antenna is used, the generated THz signal and the beat frequency signal used to generate it are synchronized at the detecting antenna resulting to the generation of THz fields having a THz frequency equal to that of the beat signal. In comparison to thermal detection, homodyne detection provides: high sensitivity since the background signals are suppressed at the receiver, they offer a good SNR and operate at ambient temperature. In addition to that, homodyne detection principle allow the THz field amplitudes and phase to be measured unlike intensity detection by thermal detectors [9]. In most cases, homodyne detection is performed either with photoconductive antennas or by electro-optic sampling:

Photoconductive detection

The initiation of ultrafast lasers in the 1980s marked the beginning of technological revolution in THz generation and detection. In 1984, Auston and his colleagues demonstrated THz generation and detection with photoconducting switches [40]. The detection of THz radiation with photoconductive switches is based on an optical pulse signal and the THz signal synchronized at the same time at a photoconductive switch (PCs). In this detection, unbiased PCs is illuminated with THz signal and probe signal, charge carriers are generated with probe signal on a semiconductor surface of the photoconductive antenna. While, the THz field accelerates the carriers resulting to the flow of a photocurrent. The generated photocurrent amplitude corresponds to the incident THz signal amplitude. The measurement with this approach preserves the phase information and the amplitude information of detected signal [12]. A signal can be sampled point by point with probe pulse to obtain the whole waveform of THz wave. This is achieved when the THz signal and the probe signal are synchronized at the switch at the same time, by scanning the delay stage over time. The coherent nature of this detection approach makes it to demonstrate a comparatively good SNR and high sensitivity of the THz signal. This is possible has it acts as a switch and the switch is only gated when both the THz signal and the probe signal are received at the switch simulteneously. Photoconductive detectors have a small volume when compared with Electro-optic detection (EO detection), photoconductive antennas do not require additional optics when compared with EO detection, therefore they offer a very promising compact THz detection scheme [41]. This detection approach could as well be applied for continous wave THz signal detection. More details on this detection approach will be discussed in chapter 5.

Electro-optic detection

Electro optic (EO) sampling on the other hand measures the electric fields of the THz pulses providing phase as well as amplitude information. EO sampling is based on the probe laser pulse and the THz pulse which are simulteneously propagated through a thin EO crystal inducing pockel effect. [42, 43]. The laser probing beam interacts with THz signal in the crystal resulting to the changes in polarization state of the probe signal which is measured with a fast photodiode [36]. The detected changes will be proportional to the incident THz fields. EO detection offers broad frequency response and high sensitivity and this is dependent on the type of crystal used and the crystal thickness. Thin crystals such as ZnTe with good phase matching properties are preferably commonly used materials for EO crystals for THz EO detection [44]. When a linearly polarized probe signal which is synchronized THz

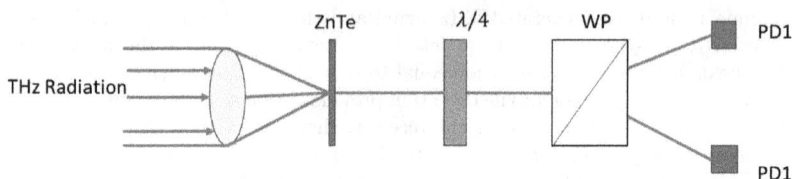

Figure 3.3: The detection of THz radiation based on electro-optics sampling employing ZnTe crystal with balanced photodiodes.

pulse, is transmitted through a crystal, the temporarily induced birefringence in the crystal changes its linearly polarized properties into a slightly elliptical properties. A polarizing beam splitter (or a Wollaston prism), splits the signal into linearly s-polarized signal and p-polarized signals which are detected by the two balanced photodiodes as depicted in Figure 3.3. An intensity difference between the two signal corresponds to the THz field [45]. The entire EO sampling of the THz transient can be obtained similar to photoconductive detection by sampling the THz signal point by point. EO detection method exhibit high spectral resolution which can only be limited by the probe pulse sampling duration [46]. Further more, EO detection requires materials with large electro-optic coefficient such as $LiTaO_3$. The high permitivity of $LiTaO_3$ generates several reflections in the crystal material. Such reflections are problematic when sampling the electric fields. Organic crystals which are alternative materials which can be used exhibit reduced permitivity and big losses at THz frequencies [41]. A crystal which is much thicker will bring an interaction which will be longer, yet on the contrary, will result in a reduction in the frequency response mainly because the group velocity mismatch of the phase velocity of THz signal and that of the probe signal [47]. In addition to that, the complexity of electro optic sampling detection techniques requiring optical setup with tight beam alignment is challenging to implement, making them commercially unappealing for emerging application demanding compact setup. In the following chapter we discuss semiconductor diode based detection.

Chapter 4

Semiconductor diode laser based THz detection

4.1 Introduction

Semiconductor diode laser-based THz emitters have proved to be compact, economical, and highly efficient [8]. Employing semiconductor diode lasers to detect THz radiation would contribute significantly towards development of extremely compact and economical THz systems. Brenner et al., demonstrated an approach to detect THz radiation utilizing inexpensive semiconductor diode laser operating at room temperature [3,4]. In their approach, they injected the THz radiation to the active region of the diode and determined the voltage variation over the pn-junction, with Golay cell used to take a reference measurement. The injected THz radiation caused a change in voltage which was detected with a lock-in amplifier. The plot of the detected voltage against the THz power monitored with the Golay cell showed a linear correlation, which depicted the detection of the incident radiation [4]. In this study, THz gas laser emitting 2.52 THz frequency radiation was used and had an average power in the range of 10-50 mW provided by pumping methanol (CH_2OH) in a homemade cavity using CO_2 laser operating at 9.6 μm emission line [3]. Further, they verified their investigation with a theoretical analysis, which demonstrated that the free carrier dynamics were involved in the detection. The free carriers absorbed the incident THz radiation energy. Therefore, heating a crystal by absorbing energy by carrier phonon interaction, the heating caused a variation in splitting of the Quasi-Fermi level which was then determined, and the observed response was compared to those of Golay cell [4]. In both cases the signal response is observed to drop greatly at modulation frequencies above 1 kHz. The slow dynamics observed represents a typical picture of thermal detectors and Brenner et al., concluded that the heating up of the semiconductor crystal was one of the vital component of THz detection mechanism [8]. We attempted to extend this investigation of THz detection with the research of the interaction of THz frequency from various sources with semiconductor diode laser based on nonlinear mixing of the signals at the p-n junction of the diode laser, however the investigation was not successful.

In this study, we focused on optoelectronic nonlinear frequency mixing of THz radiation in a semiconductor diode laser. Nonlinear optics have been demonstrated in wavelength shifting in optical communication and THz up conversion with quantum cascade laser [49] and more recently on THz detection based on resonant tunneling

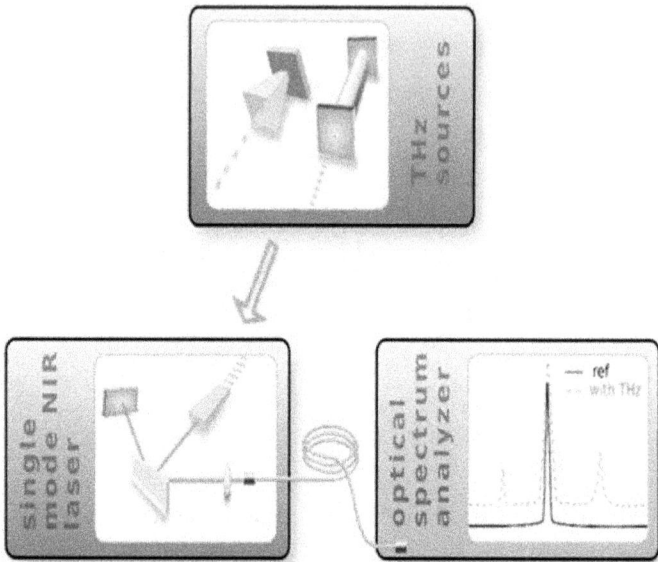

Figure 4.1: proof of principle for THz radiation conversion to NIR by injecting intense THz source into semiconductor laser arranged in Littmann cavity configuration.

laser [50]. Nonlinear carrier dynamics in semiconductor devices interacts with THz radiation, thereby shifting the input THz frequency to near infrared (NIR) by generating sidebands in NIR frequency, and could permits the transfer of long wavelength radiation to a wave with much shorter wavelength, which can be amplified and easily analyzed by means of simple spectrometer [51,52]. A proof of principle idea for THz wave convertion to NIR is shown in Figure 4.1. In this study, we investigated the nonlinear process in a semiconductor diode upon injecting intense THz signal into the active region of the semiconductor diode laser. The nonlinear susceptibility of GaAs in the active region of the semiconductor laser was expected to permit the injected THz fields to interact with the near infrared beam generating sidebands on both sides of the near infrared radiation. Thereby shifting the input THz wavelength to NIR wavelength. The approach was expected to preserve the properties of the injected THz radiation such as amplitude in the intermediate frequency. We performed a series of experiments with different THz sources, unfortunately, we didn't manage to demonstrate the nonlinear frequency conversion of THz radiation to near Infrared as was expected with compact semiconductor diode. If this were to be successful, a new inexpensive, simple, cost effective and compact room temperature THz spectrometer could have been demonstrated since the distance between the emission line and the sidebands equals the incident THz frequency. An overview on the failed attempts to obtain the expected results will be given later in this chapter.

4.2   Non-linear mixing process

Since the carrier density in a diode laser responds to THz frequencies [3], we expected that it also could have responded directly to THz radiation injected into the diode laser. This could have resulted in a THz modulation of the optical gain in the laser and generate sidebands in the NIR optical spectra, very similar to four-wave mixing experiments [48]. Thus, this effect would have enabled to transfer an injected THz signal into the NIR spectral regime where it could have been analyzed with a simple spectrometer. Nonlinear frequency conversion is generally based on the optical heterodyning process of multiplying two sinusoidal signals in a nonlinear device such as a diode laser, to generate intermediate frequencies from which the properties of the input signal can be obtained. Consider two sinusoidal signals mixed in a local oscillator with their equations given by

$$v_{THz}(t) = A \cos(\omega_{THz} t + \phi)$$
$$v_{nir}(t) = A_{nir} \cos(\omega_{nir} t) \tag{4.1}$$

where $v_{THz}(t)$ is the injected THz signal into the diode laser and $v_{nir}(t)$ is the NIR signal of semiconductor diode. The nonlinear mixing of the two signal is based on the multiplication of two signals in accordance to the trigonometric identity

$$cos(A + B) = \cos(A)\cos(B) - \sin(A)sin(B) \tag{4.2}$$

applying the identity, the output signal in this case common referred as the intermediate frequency will be the product of the two signals

$$v_{out} = v_{THz} v_{nir} \tag{4.3}$$

substituting equation (4.1)into equation (4.3) we obtain the following expression

$$v_{out} = A.A_{nir}[cos(\omega_{THz} t + \phi + \omega_{nir} t)$$
$$- \frac{1}{2}[cos(\omega_{THz} t + \phi - \omega_{nir} t) - cos(\omega_{THz} t + \phi + \omega_{nir} t)]] \tag{4.4}$$

By regrouping equal terms we have the simplified expression as

$$v_{out} = \frac{A.A_{nir}}{2}[cos(\omega_{nir} + \omega_{THz})t + \phi) + \cos(\omega_{nir} - \omega_{THz})t + \phi)] \tag{4.5}$$

The output consists of the sidebands with frequencies as the sum frequency on the upper side

$$\omega_{nir} + \omega_{THz} \tag{4.6}$$

and difference frequency of the original signal frequencies on the lower side to the NIR signal

$$\omega_{nir} - \omega_{THz} \tag{4.7}$$

equation (6) and (7) shows up conversion term and down conversion term respectively which indicates the side bands of the generated frequency due to non-linear mixing. This way nonlinear mixing of THz signal with diode signal is expected therefore to transfer of THz radiation to NIR radiation and demonstrate a compact detection scheme using semiconductor diode lasers.

4.3 Experimental description

We employed an external cavity diode laser (ECDL) with a Littman cavity design [53] as shown in Figure 4.2, emitting at 840 nm center wavelength as a nonlinear mixing medium. The laser was 1000 μm long, single section ridge waveguide laser with 5 μm width on a C-Mount from (Ferdinand Braun Institut). The diode laser had an antireflection coating (AR) on the front facet while the back facet had an high reflective (HR) coating. The diode was mounted on aluminum mount with a fixed beam height. The diode emission was collimated onto the diffraction grating with an aspheric collimation lens (focal point=6.24 mm, NA=0.40, Thorlabs, A110TM-B) which was mounted on adjustable collimation package tube attached to the diode mount. A diffraction grating which was an holographically recorded sinusoidal gold coated grating, with 1800 $grooves/mm$ on a $30 \times 19 \times 6$ mm^3 substrate from (Carl Zeiss SMS GmbH) with maximum efficiency of about 80% optimized at 800 nm wavelength was used for wavelength tuning and for coupling the signal output of the laser cavity. The grating was fixed in a mount which provided horizontal beam angle adjustment. The end mirror for optical feedback was mounted on ultra clear edge mirror mount with (3), 100 TPI locable allen keys from(Newport). The mount was stable against vibration and provide fine tuning of the wavelength with allen keys adjustment in horizontal or vertical geometry. The ECDL was mounted on a stable and compact aluminum base plate of thickness of 5 mm and operated at room temperature. With a grating mounted at 70° incidence angle, the first-order beam was reflected to the grating and back to the diode laser by a highly reflective end mirror for spectral tuning which enabled the selection of a single wavelength emission of the diode laser. Adjusting the end mirror angle the wavelength was tuned, while the position of the diffraction grating and the zeroth-order beam (output signal) remained unchanged. This is significant in employing such a configuration in real world application. A prototype of the assembled setup is shown in Figure 4.3.

In this study, Littmann metcalf configuration was used to ensure single mode output emission is maintained. This was necessary to ensure that no side modes are superimposed to the nonlinear mixing signal. The NIR radiation of the diode laser was then coupled out from the 0th order reflection at the intra cavity diffraction grating and analyzed with an optical spectrum analyzer. The laser diode was biased slightly above threshold at 50 mA. The NIR output signal of ECDL was then coupled into a single mode fiber, with 700 μW of power coupled for analysis using fiberport collimation package (Thorlabs). For our proof of principle experiment, we used three different THz sources: a SynView THz source with frequency oscillating between 230-320 GHz, a Gunn diode emitting at 120 GHz with 30 mW of power and a THz TDS system operating at 0.5 to 2.5 THz frequency range.

4.3.1 SynView THz source

A SynView THz source operating between 230 GHz - 320 GHz THz frequency with an output power of 60 μW was used to interact terahertz radiation with ECDL diode laser. Its frequency sweep time is 240 μs with an operation bandwidth of 90 GHz (230 GHz -320 GHz). More features of this source can be found in ref [54]. For the experiment, the first approach was to investigate detection of the incident THz signal by measuring the voltage variation through the p-n junction of the diode

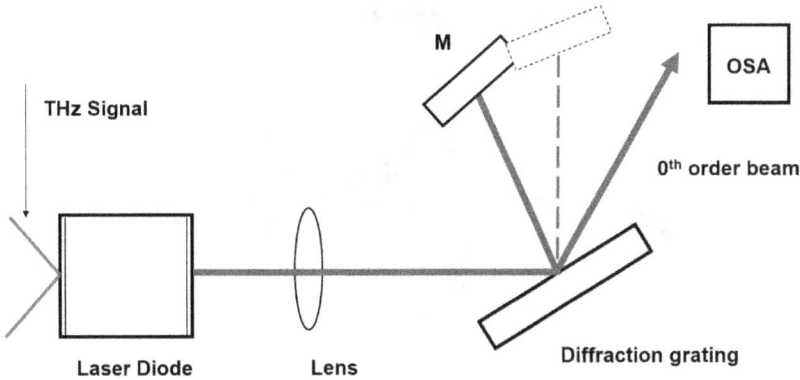

Figure 4.2: The semiconductor diode laser arranged in Littmann cavity configuration.

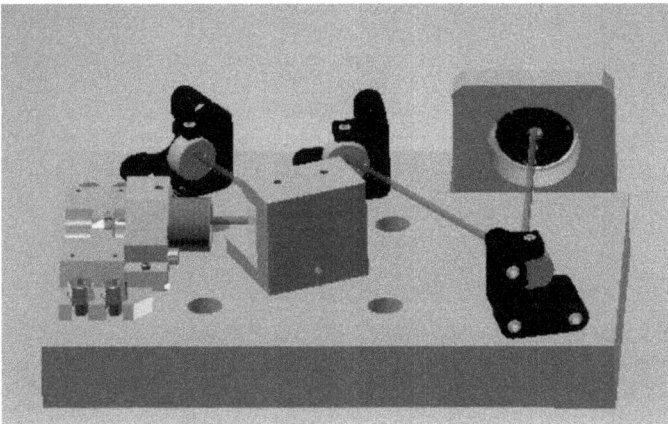

Figure 4.3: Prototype of the setup in a compact alluminium base plate with footprint of 15 cm by 12 cm

laser with a lock-in amplifier upon THz injection. In this case, the THz radiation was chopped to increase the sensitivity for the lock-in amplifier detection. Second approach was to analyze the NIR emission of the diode laser for sidebands of its central mode separated by the injected THz frequency using an optical spectrum analyzer. The polymethylpentene (TPX) lens, with 50 mm focal length was used to collect and focus THz radiation injected into the uncoated laser diode facet side opposite to the external cavity. The resultant intermediate signal was analyzed with an optical spectrometer with a resolution of $0.01nm$ and a sweep time of $20ms$, with no averaging for the sideband generation. We expected the non-linear mixing in the p-n junction of the diode laser to induce sidebands in NIR. Unfortunately, for both the experiments we could not see any effect upon injecting THz radiation on the p-n

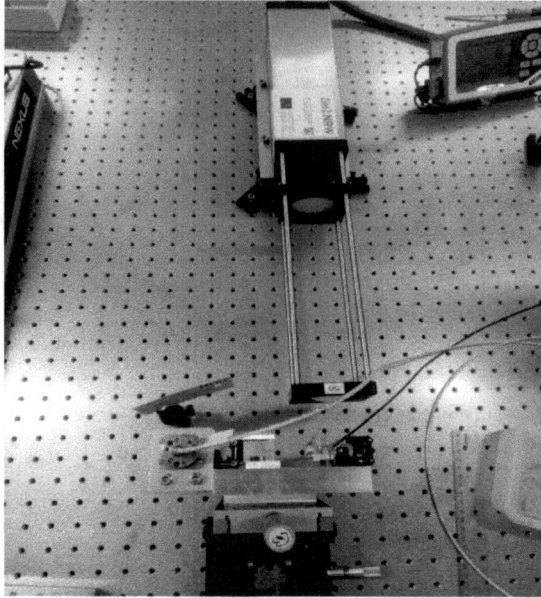

Figure 4.4: Interaction experiment with SynView Head THz source

junction of the diode laser. We later modified the injection direction of THz radiation as shown in Figure 4.4 to try to maximize the power coupled into the diode, but still not much was achieved for the interaction experiment with the SynView THz source. In the first approach, the noise was predominant on the detected signal. Upon blocking the THz signal, we discovered we were detecting only noise. Hence no interaction was observed. Similarly, in the analysis of the central mode no side modes where observed. We presumed that since the Synview THz source delivered only 60 μW of power it could have been insufficient for the nonlinear mixing effect to be noticed. Also, as the THz beam spot size was around 1 mm we thought it could have been that we were not getting enough power into the p-n junction of the diode due to the large focus of the beam hence less coupled THz power into the diode.

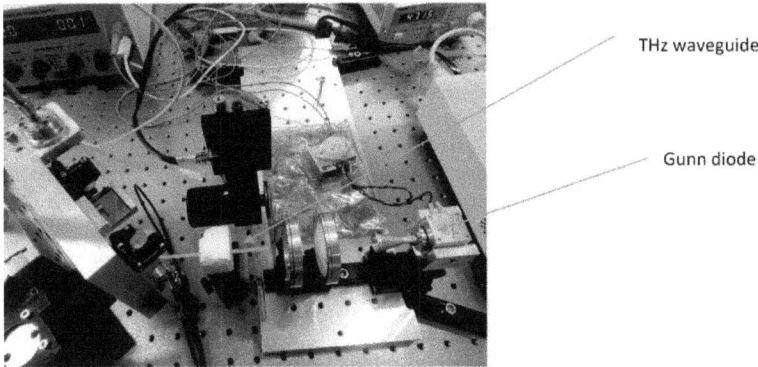

THz waveguide

Gunn diode

Figure 4.5: Interaction experiment with Gunn diode as THz source with THz waveguide used for coupling THz signal into the diode laser

4.3.2 Gunn diode

Due to the above challenges, a Gunn diode was borrowed from the University of Marburg. This THz source emits THz radiation with a frequency of 120 GHz but with even much higher emission power of around 30 mW. The experiments which were done with the SynView THz system were once again repeated with this system. Due to coupling of signal challenges, we employed THz waveguide to couple the THz signal from the Gunn diode into the semiconductor laser diode. The waveguide could deliver intensity to a very small region but unfortunately after several measurements we could not observe any promising results. Figure 4.5 shows a picture of the interaction experiment with a Gunn diode with THz waveguide used to couple THz signal into the diode.

4.3.3 THz-TDS system

Another experiment in collaboration with terahertz research group of Prof. Dr. Martin Koch in Marburg was carried out employing a THz TDS system. The THz TDS system generate THz radiation based on optical rectification in a Lithium niobate $(LiNbO_3)$ crystal that is cut at an angle of 63°. THz radiation is generated using a technique often referred to as "tilted-pulse-front pumping scheme" similar to the one described by Hirori et al., [55]. The pump laser used was an amplifier laser system (spitfire Ace-35F)from spectra physics which is an advanced form of Ti:Sapphire laser system. The system provides very strong 35 fs intense optical pulses at a 800 nm central wavelength with a repetition rate of 1 kHz at an average power of 7 W. The schematic for THz generation using this system is shown in Figure 4.6. Standard achromatic doublet lenses are used between the grating and the crystal instead of the cylindrical lenses as described in Hirori work. The generated THz pulsed signal delivers an average output power of ~ 1 mW, its beam spot size (FWHM) is 900 μm and covers a broad frequency of range from 0.5 to 2.5 THz.

The generated THz frequency was focused into the diode p-n junction using the parabolic mirrors, with the diode laser inserted at the focal point of the parabolic mirror. The resultant intermediate signal(THz signal plus semiconductor signal) was then focused to the fast photodiode with a rise time of 3 ns (Newport 1807-FS) and a Stanford Research SR830 Lock-in amplifier was used for sensitive detection and it was locked into 1 kHz frequency. A digital oscilloscope (Tektronix) with a bandwidth of 200 MHz and a rise time of < 2.1 ns was used for the display of the signal. Figure 4.7 shows the schematic of the arrangement of the experimental setup we used to interact semiconductor signal with THz TDS source signal in Marburg. The intermediate signal was coupled into a single mode fiber and then focused into a fast photodiode with an oscilloscope displaying the signal analyzed. For sensitive detection the oscilloscope in Figure 4.7 was replaced with a lock-in amplifier by directly connecting a photodiode to it. We expected upon injection of a strong THz pulse into an operating laser diode, carrier dynamics and nonlinear mixing of THz signal with laser diode signal to take place. This could result to the intensity dynamics and our expectations was to observe a drop and fast recovery of the laser signal, faster in μs, otherwise thermal effect. The second observation would have been the appearance of new spectral components in optical spectrum alongside the main lasing mode due to nonlinear mixing which could have been similar to a four wave mixing experiment, thereby generating sidebands. Unfortunately, we could not observe any of the expected results as we encountered the instability problems of our diode signal. The fluctuations came about due to the background radiations which were stronger than the diode signal. The scattered pump signal plus electronic noise from the other instruments in the lab contributed to this background noise which was picked by the diode current controller and resulted to the signal flunctuating from the expected signal base line.

4.4   Results and discussion

4.4.1   Diode characterization

Figure 4.8 illustrates the current power characteristics of the ECDL used in the investigation of this study. It shows the power detected when the first order beam was fedback to the diode and when it was blocked off with measurement taken with Thorlabs power meter (P100) at room temperature. The threshold current of the diode was 50 mA. The non-linearity of the I-V characteristics curve makes this diode appropriate as a nonlinear device for THz detection via nonlinear frequency mixing. The measured spectrum Figure 4.9 shows the single mode emission at a wavelength of 840.2 nm corresponding to 358 THz frequency. The emission of the diode at fixed grating position was tuned by adjusting the end mirror. Figure 4.10 depicts a color map of two modes 837.2 nm and 838.1 nm obtained by adjusting the end mirror. This was done to check that the laser operates completely in single mode emission to ensure that a clean signal was involved in the nonlinear mixing experiments hence no competing mode induces the sidebands. By injecting intense THz radiation into the active region of the diode laser, our expectation was that it interacts with near infrared radiation resulting in the generation of sidebands on both sides of the main NIR signal. The NIR side bands could then be collected alongside the NIR carrier with a single mode fiber and analyzed with an optical spectrum analzyer (OSA).

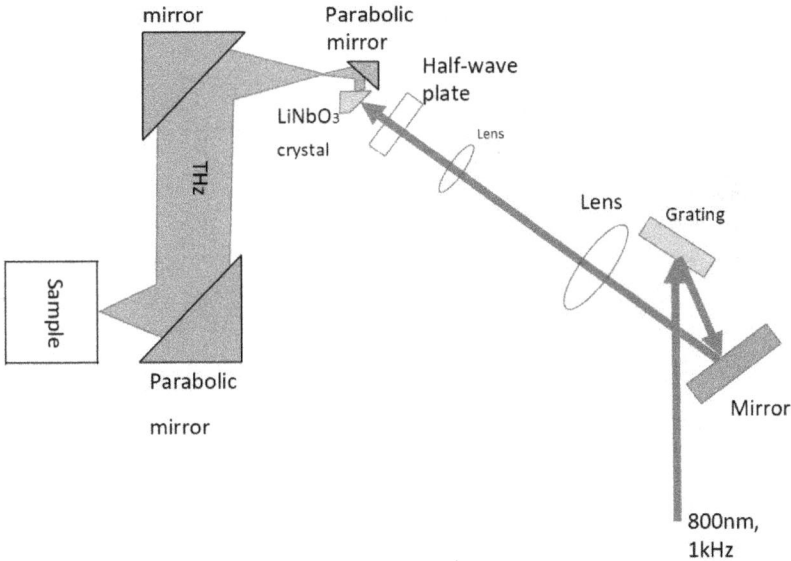

Figure 4.6: Schematic drawing of the THz TDS system setup in Marburg for THz generation via tilted-pulse-front pumping scheme. Drawing courtesy of Philipp Richter.

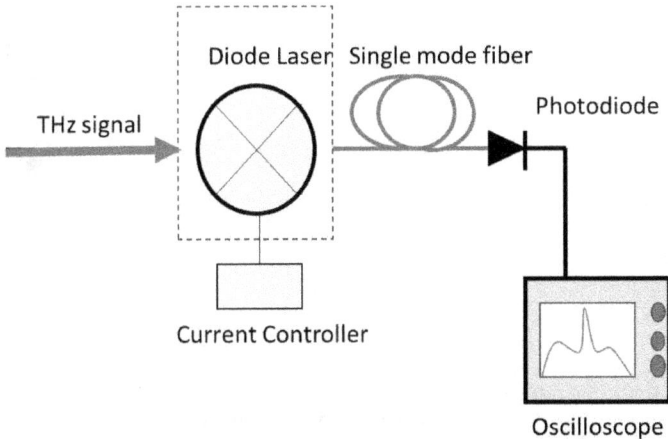

Figure 4.7: Schematic of the interaction experiment using a THz TDS system THz signal and balanced photodiode as a detector for analysis of the signal.

Unfortunately, the different measurements we did with the SynView source as well as the Gunn diode were unsuccessful as we did not manage to observe the expected

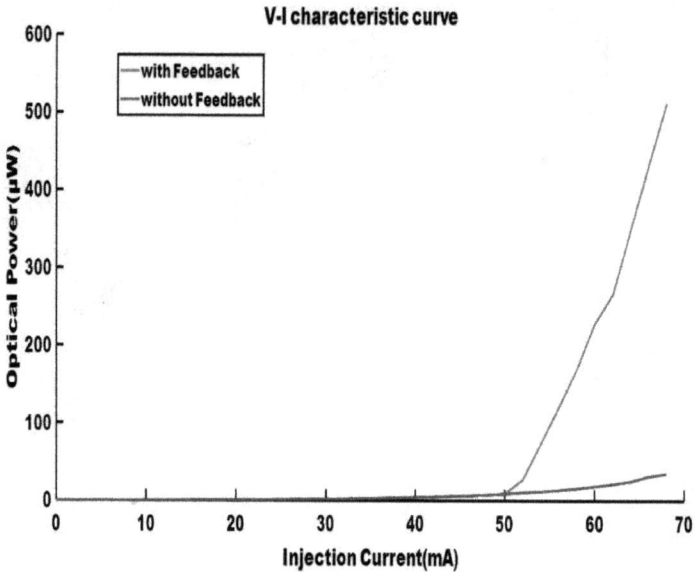

Figure 4.8: VI characteristics of the diode with the laser operated in continous mode at room temperature.

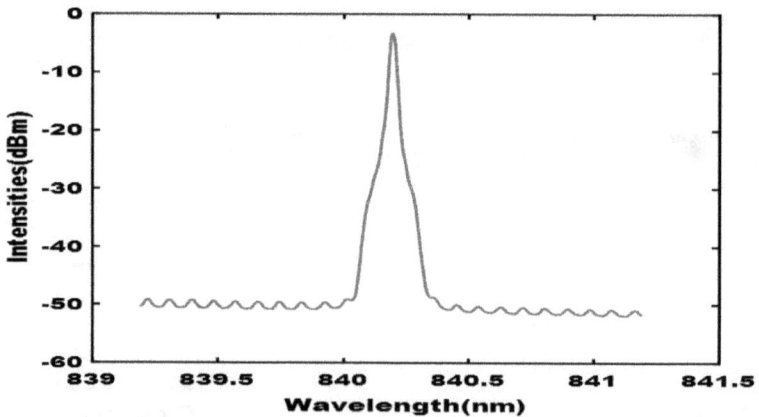

Figure 4.9: Optical spectrum of the diode laser obtained with optical spectrum analyzer with a resoluton of 0.01 nm.

results. Figure 4.11 illustrates the expected appearance of the sidebands alongside the main mode for the different THz sources.

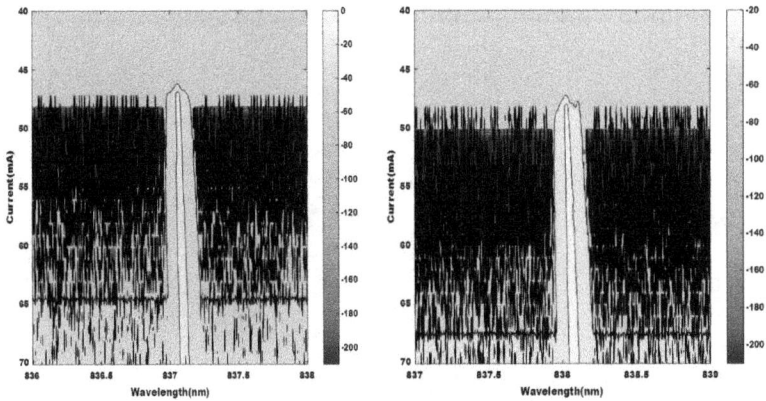

Figure 4.10: Color map of two different modes tunable by adjusting the end mirror.

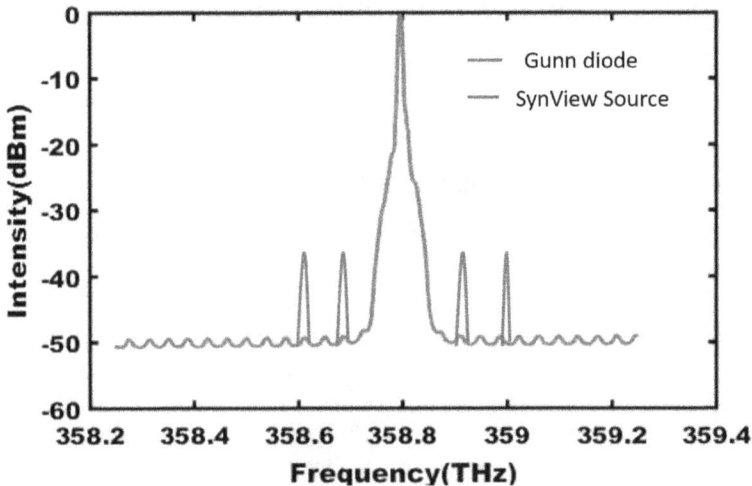

Figure 4.11: Optical spectrum with the expected sidebands alongside main mode for 120 GHz Gunn diode and 230 GHz SynView THz source.

Another experiment was perfomed in University of Marburg with the THz TDS system and a fast photodiode with a rise time of 3 ns used to evaluate the temporal behaviour of signal upon THz injection. Here, we expected that upon THz injection, carrier dynamics and nonlinear mixing with laser mode to take place. Our expectations was to observe a drop in the intensity and fast recovery of the laser signal in the process this dynamics was expected to increase the lasing threshold. Thereby photon emission by the diode could be temporarily interrupted resulting to sudden drop in power before the diode regains the charge carriers and the lasing

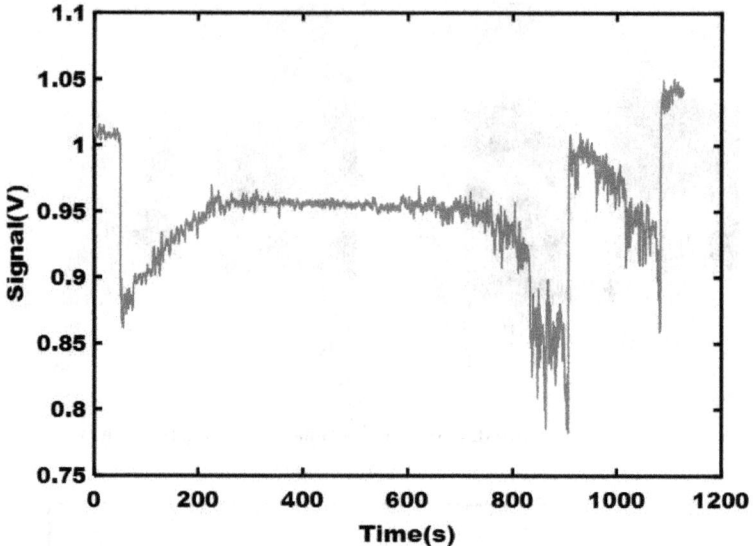

Figure 4.12: Output signal measurement of the diode laser with lock in amplifier without THz signal input.

process resumes again. Unfortunately, during measurement of this experiment the diode developed unprecedent noise which made the diode output signal to fluctuate up and down even before any THz signal was injected. The diode signal instability were also observed upon THz injection.

4.4.2 Interaction experiment with THz TDS system

To quantify the diode stability, the output signal of the diode laser which was focused into the photodiode was chopped and was measured as a function of time without THz signal with a lock-in amplifier (LIA). The diode output was seen to fluctuate up and down from time to time due to unexpected noise. To lower the noise as good as possible a LIA was employed to ehnance the sensitivity of the detection. The observation of this signal fluctuating from time to time was not periodic and in between the fluctuations the signals also increases or decreases as shown in Figure 4.12. However, we then tried to see if the THz field induces a change to the diode output when operating the diode laser near threshold regime. Therefore, the THz TDS signal was chopped at 50 Hz frequency. The results are shown in the Figure 4.13. All data points arise from measurement taken for 30 minutes and sampled after every 100 ms. If the THz was blocked, we expected to see nothing. Therefore, the black dots named "THz blocked" was the noise level and this could serve as the reference signal. Additionally, we measured both channels of the lock-in amplifier to make sure we do not miss any signal in the out-of-phase channel 2. When adding the THz signal to diode signal a boxcar was used to restrict the measurement to a

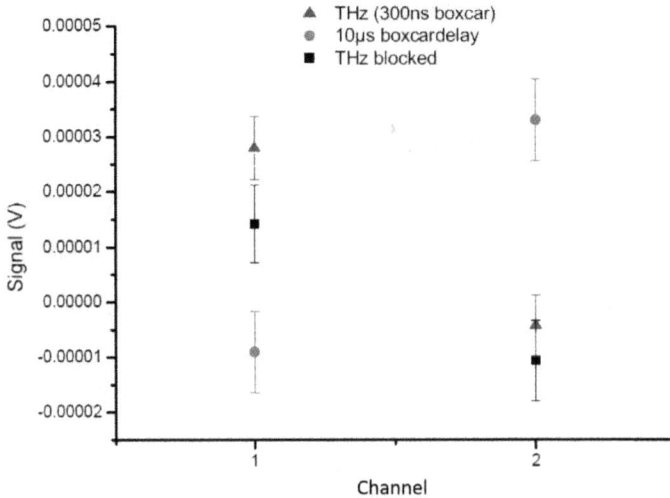

Figure 4.13: Interaction signal measurement when THz signal is blocked and when not blocked with two lock in amplifier channels.

period of 300 ns within the arrival of the THz-pulse to further lower the noise contribution. The results are shown in blue. There was a slight increase in channel 1 signal while the out-of-phase signal remains unchanged within error of measurement. The third measurement (red) shows the signal when the boxcar window of 300 ns was delayed by 10 μs with respect to the THz-pulse striking the diode. So, in this case we expected the THz-induced distortion to have vanished and, therefore, we should measure the reference signal again. However, the red signal was drastically different from the reference (black) and the THz (blue) signal. The uncertainty and unreliability which could have been introduced in the measurement due to the fluctuations demanded that the diode be stabilized before further investigations could be carried out.

4.5 Conclusions and recommendations

In summary, we have discussed the study of the interaction of a THz signal from various sources with a semiconductor diode laser, with particular focus on investigation to use diode to detect THz radiation. Our expectation was that the detection would have taken place in the carrier plasma. Since the charge carrier in a diode laser responds to THz frequencies, we hoped that it would also respond to THz radiation injected into the diode laser. As a result THz modulation of the optical gain in the laser was expected to occur and generate sidebands in the NIR optical spectra. Thus, the injected THz signal would have been converted into the NIR spectral regime and could be easily analyzed with an optical spectrometer. Unfortunately, we didn't observe this for various reasons:

- The SynView THz source: emits around 60 μW of optical power, since this amount is low chances are that the amount of power that could have been delivered into the diode for interaction experiment would be significantly have reduced. This could be due to losses which might have occured during the injection of THz signal into the p-n junction, hence chances are that the amount of signal coupled into the diode could have been insufficient for any chances of interaction between THz signal and diode signal to take place. Another drawback could have been the large beam size of the THz signal ~ 1 mm which was challenging to focus into small spot in the diode laser. Due to the fact that the SynView system scan frequencies in the range 230 GHz - 320 GHz, the rapid change of frequencies may make it difficult to observe the sideband since before the sidebands are generated they may be cancelled by the next operation frequency.

- Gunn diode: despite the fact it emitted more power, 30 mW, the beam size of THz signal at lower frequencies is too broad hence coupling into the diode active region was quite challenging. We used THz waveguide but still it seemed we didn't couple enough signal power to the diode to be able to observe any expected interactions results.

- THz TDs system: we experienced some instability of our diode laser which made the signal to fluctuate during the experiment. We tried to suppress the noise in the measurement by employing lock in detection. However, the noise was always there hence it become challenging to take further measurements. A part from that, there was a likelihood that the interaction of THz pulse signal with diode laser signal might have occurred at a very short time interval hence unnoticed, this was due to the fact that we used pulsed signal while the operation of the diode was in continuous mode. But due to the noise fluctuations all this could not be noticed.

We recommend future work to be done with the hope to demonstrate that indeed this idea could still be viable for THz detection. Powerful THz sources should be used and the following to be taken into consideration.

- CW THz source: powerful THz source such as Methanol gas laser generating 2.52 THz with powers in the range 10-50 mW to be used similar to one employed by Brenner et al [8], or any other CW source which can deliver intense THz signal power. In this case the laser should operated in CW mode. Silicon lenses should be incorporated in the coupling process of the injection of the THz radiation. This will increase the possibility of ensuring more intense power reaches the p-n junction.

- Pulsed THz source: a powerful pulsed source should be used, however, the diode should be temperature stabilized and if possible shielded from any random noise interference of the signal from the surrounding instruments. In this case the diode laser injection current should be modulated to match the repetition rate of the pulsed THz source.

We recommend future investigation to be carried out with the operation of external cavity diode laser operated above threshold and THz pulses with 1 kHz repetition frequency injected into the diode.The following series of experiments remains can be performed: First investigation could be to observe the diode laser emission with lock-in-amplifier locked to repetition rate of amplifier. Second attempt could be to observe the diode laser emission with fast photodiode and oscilloscope (any dynamics faster than microsecond range. Up conversion of diode laser emission with NIR amplifier signal for best time resolution (picosecond transients expected in best case). Third investigation could be to observe the diode laser emission with OSA locked to 1 kHz repetition frequency of the injected THz pulses. Finally one can try to observe the diode laser emission with spectrometer and lock-in amplifier with THz spectral filtering (e.g. 1 THz) done to search for corresponding sidebands in diode laser spectra. The experiments perfomed here were done by injecting the THz radiation directly to the rear facet of the diode, one could also try to change the input direction to see if there is any effect. Another factor we didn't take into consideration during this investigation was whether the beam polarization could have any effect on the results. However since polarization of both the THz and diode signal may be different, one should consider investigating if the way the beams are oriented may influence the overal mixing of the signals in the diode.

Chapter 5

Monolithically integrated two-color Distributed Bragg Reflector semiconductor laser diodes for THz emission

In this chapter, we characterized two monolithic two-color diode laser strcutures and demonstrated their potential application in building compact and cost effective CW THz systems. The first device enables the adjustment of laser modes through changing of the laser temperature plus that of the diode submount simultaneously, and the second device tunability was achieved via current injection to the laser chip alone. We discuss in this chapter, the potential application of the developed systems for non-destructive moisture measurements, spectroscopy and THz metrology.

5.1   Introduction

Optoelectronic techniques are driving the acceleration of the transfer of THz technology out of the lab setting into the real market application, through the development of robust systems meeting industrial demands. Over the past decade, THz TDS as developed into a useful tool for THz research. Its unique capabilities have enabled many interesting properties of materials and substances to be accessed and revealed ranging from intra and intermolecular vibrations [56, 57], carrier dynamics in semiconductors [46], as well as rotational transition of gas molecules [58]. A complete THz electric field signal generated from THz TDS system can enable the phase and amplitude information of the sample to be accessed with broad frequency spectrum. This has enabled this systems to gain a broad scope of applications in many fields [56, 58]. Nevertheless, the use of ultrashort lasers such as Ti:Sapphire laser to drive ultrafast photoconductive antennas or electro-optical crystals in THz TDS system to generate broadband THz frequency with high intensities power at average repetition rate [59], is complex and results in a bulky system which will demand high investment costs to implement. This has hindered the wide adoption of THz TDS systems for scientific and industrial applications. On the other hand, in terms of the attainable spectral resolution CW THz sources exhibits superior features when compared to broadband pulsed emitters. This is possible when generated with a pump laser with a narrow linewidth resulting in a THz signal with narrow linewidth. The narrow linewidth ensures that power distribution is concetrated on a single frequency thereby a higher spectral power density is observed in CW THz signal unlike in pulsed THz system where power distribution is on the entire THz spectrum [60, 61].

In principle, the narrow linewidth is expected to enhance the SNR significantly as the signal will have high power density which can penetrate a large depth of the sample which is desired for spectroscopy or imaging where noticeable characteristics can be revealed at narrow frequency range [61]. Applications that require to sense multiple gas mixtures, characterize different materials and perform imaging at specific frequencies demands a source that emits at specific single THz frequency, in such application broadband pulsed terahertz frequency sources may become unsuitable and undesired. Widely tunable CW THz emitters over the entire THz spectrum range are attractive for such application, this is due to different molecules of materials having pronounced absorption spectral finger prints at specific THz frequency [59]. CW THz emitter systems that are compact, cost-effective, emitting radiation with high spectral purity and high power, and tunable over a wide wavelength range are scarce.

THz photomixing in combination with compact diode lasers promises the realization of such extremely compact, affordable and effective THz sources [62, 63]. THz photomixing involves optical heterodyning of two near-infrared lasers beams with an offset in frequency in a photoconductive antenna to generate electromagnetic radiation whose frequency is same as that of the beat frequency of the exciting lasers modes [62, 64]. The generated THz wave is radiated into free space with antenna structure. A second photomixer in a homodyne setup system can be used to detect coherently the generated THz fields [65]. The coherent nature of this detection approach enables one to achieve a relatively high SNR, since the effects of background radiation are highly suppressed at the receiver [66]. The receiving antenna acts as a switch and is only gated when the beams arrive at the photoconductive antenna gap at the same time.

The spectral linewidth of the emitted THz signal depends on the exciting pump laser linewidth, therefore photomixing systems can generate THz radiations with very high spectral resolution if the used laser linewidth is narrow [67]. The narrow linewidth exhibited with diode lasers implies narrow THz signal to be generated. Hence, due to the concentration of power on a single frequency the beam is much more homogeneous due to minimum unwanted beam artifacts as a result of aberrations and dispersion. The homogeneity of the CW THz signal ensures that the THz radiation penetration depth into a sample under test is much higher resulting in better images with higher spectral resolution [68]. In addition to narrow linewidth, wide tuning range for generated THz frequency could be achieved provided a widely tunable, two-color laser is used to excite the photomixers [36].

In this regard, monolithic two-color lasers have become versatile tools for CW THz photomixing process in comparison to the conventional two independent lasers system mainly due to their simplified setup, compactness, simplified spatial mode overlaps of the two modes, affordability and highly stable optical beat frequency [63, 69]. In the recent past, a lot of research progress in semiconductor fabrication, have yielded two-color diode lasers with even more improved optical properties. Several technical breakthroughs made in the past few years have resulted to diode laser with enhanced features such as broad wavelength tunability range [70, 71], stable emission with relative wavelength drift and low phase noise [72], higher emission optical powers [28,73], dual wavelength with ultra-narrow linewidth [74], and many more. A variety of two-color diode laser concepts such as DFB and DBR diode lasers have been used to demonstrate THz generation via photomixing [59,63,69–71]. They offer great potential in the implementating novel CW THz systems with narrow linewith, stable emission, higher resolution as well as broad wavelength tunability that is desirable for high resolution spectroscopy and imaging applications.

The motivation in this chapter, was to contribute towards the development of compact, stable and affordable THz systems with potential in real market applications. In this regard, we demonstrated a CW THz system using two-color DBR laser diode as excitation source. Photoconductive antennas based on ion-implanted gallium Arsenide (GaAs) logarithmic spiral structure were employed to transmit and receive the THz radiation. The investigated diode lasers emitted two modes simultaneously in the center wavelength of 785 nm. In the first case, a single DBR laser structure consisted of two DBR lasers which are monolithically integrated with a Y-shaped waveguide and their emission made to exit the laser cavity with a single common output. From optical characterization the optical beat frequency of the laser was found to be 296 GHz. A coherent homodyne detection scheme was used to confirm the generated THz radiation. The difference frequency obtained by changing the temperature and current was varied in the range 280 GHz to 320 GHz. Finally, we demonstrated a potential area of application of the developed THz source in non-destructive moisture measurements.

In the second case, we will discuss the implementation of a compact CW terahertz source, employing an electrically tunable monolithic distributed Bragg reflector diode laser which was designed with capability to adjust the optical beat frequency of the emitted wavelengths via carrier injecton. Optical beat tuning was achieved with micro resistor heater which was implemented on top of the DBR section. It changes the local temperature of the DBR section via current injection thereby shifting the emission wavelength as a result of the changes in the refractive index. We demonstrate the suitability of the system in CW THz wave generation with the variation of optical beat frequency in the range from 100 GHz-300 GHz. This frequency band was only limited due to mode overlap at higher resistor currents which resulted to zero wavelength difference of the two modes. Further, we demonstrated the potential use of the implemented system in THz spectrocopy in characterization of THz bandpass filter and in THz metrology for contact free measurement in thickness determination of a sample of PE block.

This chapter is organized into eleven sections: A brief overview of terahertz photoconductive antennas (PCAs) is given in section 5.2, followed by the generation of pulsed 5.3 and continuous wave THz radiation with PCAs in section 5.4 with some theoretical analysis on how the generation process takes place in the semiconductor material given in section 5.3 and 5.4 for both pulsed and continous wave systems . The properties of some widely used photoconductive substrate material is descriped in section 5.5. Different concepts for dual wavelength lasers are presented as excitation sources for photomixing in section 5.6. More focus will be on monolithic dual wavelength laser sources concept since it is the one used in the investigation of the results obtained in this dissertation. The experimental setup with CW THz source is described in section 5.7. The results are presented in section 5.8 with measurements with two color DBR laser and electrically tunable two color DBR laser

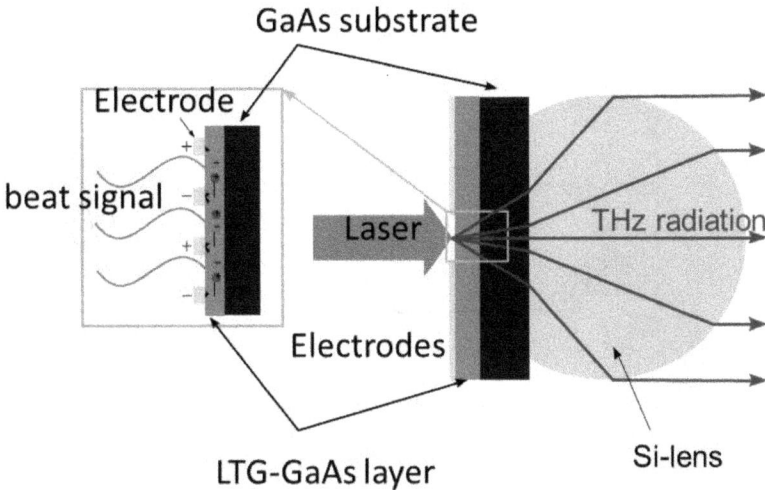

Figure 5.1: Photoconductive antenna structure with LTG-GaAs layer with electrodes on the photoconductive gap on GaAs substrate and Si-lens attached at the rear end [76].

given in sections 5.9 and 5.10 respectively with illustrative examples for potential industrial application for the developed THz system in non-destructive evaluation (NDE) of moisture measurement and THz metrology measurements for thickness and refractive index measurements. Finally, a conclusion is given in section 5.11.

5.2 Terahertz photoconductive antennas

Terahertz photoconductive antennas (PC) are electrical switches made of parallel metallic microstrip electrodes separated by a gap lithographically printed on a semi-insulating semiconductor substrate. They exploit the increase of electrical conductivity of the semiconductor material to generate broadband terahertz pulses or continuous-wave (CW) terahertz signals when excited with either femtosecond pulsed lasers or continuous wave lasers respectively. The semiconductor material, electrodes, and antenna structure form the main building blocks making up a photoconductive antenna. The semiconductor material should have a very short carrier lifetime to enable quick recombination of charge carriers once excited with an optical signal. The carriers are then accelerated by biased electrodes resulting in electromagnetic THz fields, which are coupled into free space using the antenna structure [75]. Figure 5.1 shows schematic diagram of a photoconductive antenna with all the essential elements.

Different contact geometries of electrode shapes in combination with the semiconductor substrate on which they are deposited play a significant role to radiate and detect THz radiation. These geometries vary in dimensions from a few microm-

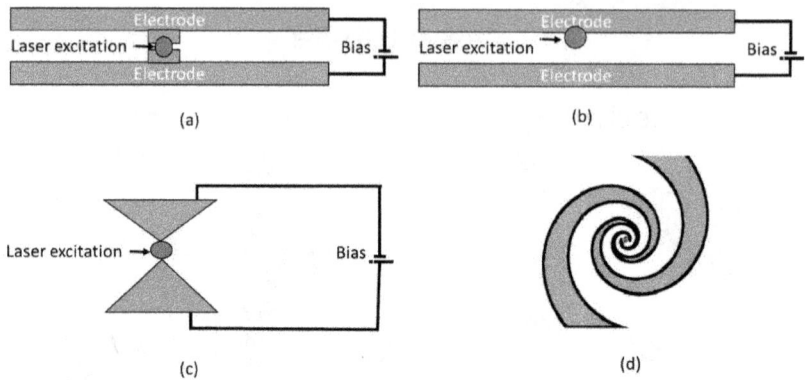

Figure 5.2: Different schematics of photoconductive antenna geometries (a) dipole (b) Stripe line(c) Bow-tie and (d) Logarithmic spiral [77].

eters in length scale to a few millimeters and govern how photoinduced currents flowing through them and the electromagnetic field, are coupled into free space at THz frequencies. Some of commonly used contact designs of PC antennas include Strip line, Bow-tie, and Dipole contact geometries on semiconductor substrate [77] and are shown in Figure 5.2 (a), (b), (c) respectively. The emitted THz radiation power intensity and spectral range can be engineered through the design geometries of PC antennas, the material of the substrate, geometry of the active area, antenna structure, as well as excitation laser parameters [78]. The PC gap area, which is the active region of the PC antennas also plays a crucial role to radiate and detect THz waves. The careful design on how the fields are distributed in the PC, will increase the breakdown threshold, thereby enabling the PC to emit sufficiently intense THz fields. PCs with small area of the gap are very sensitive, particularly when excited with low power excitation laser. However, a large gap requires to be excited with even much excitation power as well be biased with huge voltage which results in high emission of THz power [79]. In the recent years, new contact designs have been introduced, with improved photoconductive switches performance and functionalities that were not captured in the earlier designs.

One of these concepts is interdigitated metal fingers. Interdigitated metal fingers currently are one of the most popular electrode designs for CW photomixers. This design is popular since it increases the effective gap between electrodes and consequently provides a large area of interaction with the incident optical beat radiation thereby generating higher radiation intensities with broad frequency bandwidth [75]. Another established design for CW photomixing is self-complimentary logarithmic spiral antenna geometry. The photoconductive antenna with self-complimentary logarithmic spiral antenna (commonly referred log-spiral antenna) geometry design is used in the study reported in this thesis and is based on the design reported in [80]. Interesting feature of log-spiral antenna which makes them attractive for CW photomixing is their radiation pattern, impedance and their polarization properties remain unchanged in a broad spectral range [81]. Figure 5.2 (d) shows a structure of a log-spiral antenna. In some incidences the electrodes act as a waveguide of

THz radiation, along which the generated radiation could propagate. Normally, a lens which is non-absorptive and non-dispersive in the THz spectral range is used to perform beam focusing and guide THz signal out of and into the semiconductor substrate as shown in Figure 5.1. The refractive index of silicon $n = 3.47$ matches very well to that of GaAs $n = 3.6$ which ensures that the reflections between the lens and the antenna substrate are minimal [82]. Hyper-hemispherical lenses made from highly resistive silicon material are widely used since they meet the aforementioned properties and provide a better index matching with the semiconductor substrate hence efficient coupling of the generated THz wave [83].

The various antennas are classified into two categories independent on their frequency response i.e resonant and non-resonant antennas. For resonant antennas, THz emission is centered around a certain central frequency referred to as resonant frequency whose wavelength in the substrate is given by $\lambda_n = \frac{2L}{m}$, and its corresponding wavelength in free space is $\lambda = \lambda_n \times n$, where m is an interger, n represents the refractive index of the PC substrate, and L is the separation (width) between anode and the cathode and its value is constant. Dipole and stripeline antenna are examples of resonant antenna. A non-resonant antenna exhibits an adjustable width which leads to a wide frequency sensitivity range of the PC. Bowtie and logarithmic-spiral antennas are a class of non-resonant antenna geometries [79].

5.3 Terahertz pulse generation and detection in PCAs

The generation of Terahertz pulses with PCA is basically based on the ultrafast variation of the surface photoconductivity of the semiconductor substrate of the antenna under the influence of femtosecond pulse illumination. Here, a femtosecond optical pulse with a short time scale is incident on the antenna gap whose electrodes are biased, it excites the gap and propagates into the photoconductor. The concentration of electrons and holes rises sharply within a short period of time as signal is being absorbed inside the photoconductor. With the bias fields, the generated photocarriers are accelerated to produce a THz electric fields, which are radiated into free space as THz pulses. Figure 5.3 shows a schematic illustrating photoinduced generation of the charge carrier in a photoconductive gap of a LT-GaAs which is irradiated with an optical pump signal resulting to flow of a current. The flux lines indicates the generated electric field [84].

Drude–Lorentz model for charge carriers transport in semiconductors material illuminated with ultrashort laser pulses is introduced here briefly to describe the phenomenon involved for the movement of carriers during and after photoexcitation. More detailed description of this model could be found in in the following reviews [57, 85]. Consider a PCA antenna which is optically excited with ultrashort laser pulses, ultrafast changes of the surface photoconductivity of the semiconductor substrate take place with the formation of carriers in the conduction and valence bands [86]. Consequently, the is an increase in the surface conductivity of the semiconductor σ which is proportional to the charge carrier concentration.

$$\sigma = q(n_e \mu_e + n_p \mu_p) \tag{5.1}$$

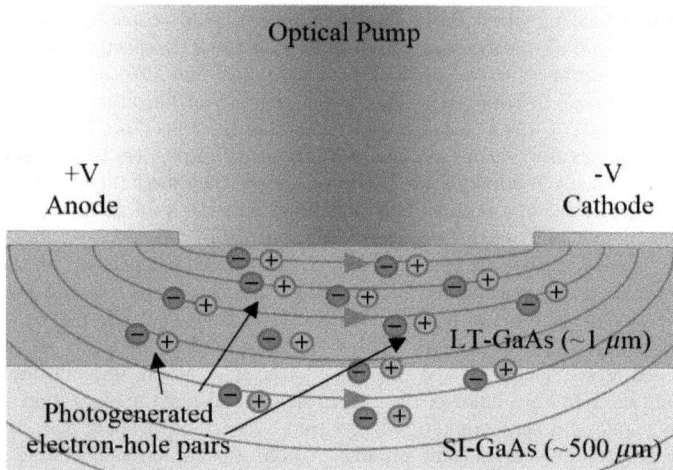

Figure 5.3: Photoinduced electron-hole pairs emission at the photoconductive gap of a LT-GaAs illuminated with an optical pump beam on PCA [84].

q is charge concentration in the semiconductor material, n_e and n_p is the density of electrons and holes while μ_e and μ_p are their corresponding mobility. The sharp increase in photogenerated charge carriers for a very short time as a result of the increase in surface conductivity of PC gap, makes the electrons and holes to accelerate in a random directions inducing the external electric field in the gap as shown in Figure 5.3. In this case, the velocity at which the carriers are moving follows differential equation given by [57,85].

$$\frac{\partial \nu_{e,p}}{\partial t} + \frac{\nu_{e,p}}{\tau_{eff}} = \frac{q_{e,p}E}{m_{e,p}} \tag{5.2}$$

where $\nu_{e,p}$ is the drift velocity, $q_{e,p}$ is the charge, $m_{e,p}$ represents the mass of electrons and holes, τ_{eff} is the effective recombination lifetime which is ~ 30fs for low temperature grown GaAs, the subscripts e and p represent electrons and holes respectively and E is the field where electrons occur and is related to the external bias field E_{bias} by [87].

$$E_{bias} = E + \frac{P}{\alpha \varepsilon_o} \tag{5.3}$$

where P is the polarization induced by the spacing of electrons and holes, α is the static dielectric susceptibility of semiconductor, ε_o is the vacuum permittivity. E is much smaller than E_{bias} due to screening effect of the space charges. The time dependence of charge polarization can be described by:

$$\frac{dP}{dt} + \frac{P}{\tau_{rec}} = J(t) \tag{5.4}$$

where τ_{rec} is the recombination time of holes and electrons, and J is the current density on the semiconductor surface described by $J(t) = q(n_e v_e + n_p v_p)$ [85].

Using maxwell's equations the above set equations can solved to determine the field of the generated THz signal. The generated field strength of THz radiation, $E_{THz}(t)$ is analogous to the time derivative of the transient current $J(t)$ given by [57]

$$E_{THz}(t) \propto \frac{\partial J(t)}{\partial t} \tag{5.5}$$

considering that the relative speed ν between an electron and a hole is provided by $\nu = \nu_h - \nu_e$ then equation 5.5 simplifies to

$$E_{THz}(t) \propto e(\nu\frac{\partial n}{\partial t} + n\frac{\partial \nu}{\partial t}) \tag{5.6}$$

with the first term representing the electromagnetic radiation due to the variation of the carrier density, second term represents the electromagnetic radiation which is related to acceleration of the electron and holes due to the electric field [86]. The THz radiation emitted by the pulsed photocurrent transient is radiated into free space by integrating the photoconductive area with an antenna.

Detection of the THz pulses generated is accomplished by optical sampling within a photoconductive antenna structure which has same properties to the structure used for THz generation. A fraction of the laser pulse used for the generation of THz signal is used in the detection process. Both the probe pulse and the generated THz signal are sent to unbiased detecting PCA antenna. The signals are synchronized to arrive at the antenna at the same time with the THz signal providing the required field (insteady of applying an external field) to accelerate the carriers and the probe pulse used to gate the antenna gap. The generated photocurrent in the antenna is a picture of the electric field received by the antenna plus the THz transient which is induced by the probe signal can be described by [57,88].

$$J(\tau)\infty \int_{-\infty}^{+\infty} E_{THz}(t)\, g(t - \tau)\, dt \tag{5.7}$$

If a photoconductivity response g(t) is considerably fast in duration than the THz field strength $E_{THz}(t)$, then the detected current is proportional to the THz field strength and the Fourier transform gives the frequency response as $J(\omega) \propto E_{THz}(\omega)$. If the photoconductivity takes much longer duration time, then the detected current will be proportional to the integral of the THz field and corresponds to a frequency response $J(\omega) \propto E_{THz}(\omega)/\omega$. If both the functions in e.q 5.7 are integrable , the Fourier transformation will yield a photocurrent with a frequency response which is described by

$$J(\omega) \propto I(\omega)R(\omega)E_{THz}(\omega) \tag{5.8}$$

where $I(\omega)$ is the pumping laser pulse spectrum, and $R(\omega)$ represents the frequency response of the charge carriers excited in the photoconductive switch [57,88]. The concept of how pulsed THz radiations are generated and detected in a PCA with a dipole antenna is illustrated in Figure 5.4.

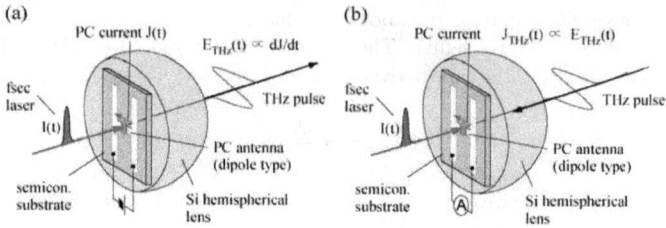

Figure 5.4: THz generation and detection of pulsed signal with PCA (a) a PC emitting antenna and (b) a PC detecting antenna [88].

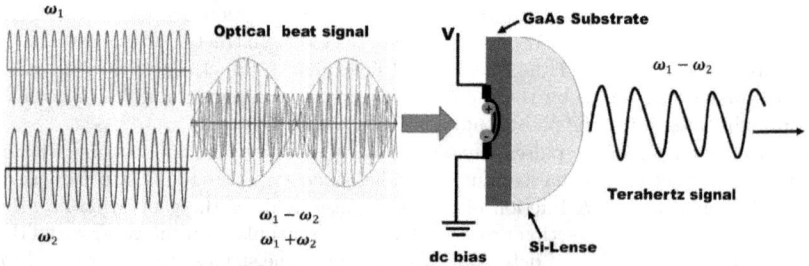

Figure 5.5: Schematic diagram illustrating CW THz generation in dipole antenna

5.4 Continuous wave THz radiation generation via photomixing

Continuous wave (CW) THz waves generation is accomplished by optical heterodyne process commonly refered to as photomixing. The process of photomixing involves optical beating of two oscillating frequencies ω_1 and ω_2 with a slight frequency offset in a photoconductive antenna results in a periodic modulation of the photoconductance of the PCAs. Together with an applied dc bias voltage at the electrodes, the optical beat induces conductivity in the semiconductor surface and creates a varying current J(t) which is radiated into free space as electromagnetic THz field with its frequency equal to the optical beat frequency $(\omega_1 - \omega_2)$. Figure 5.5 illustrates the principle of generating CW THz radiation with dipole antenna.

5.4.1 The mechanism of THz photomixing

For CW THz generation, the surfance conductivity of the semiconductor and charge carrier acceleration are similar to that of pulsed PCAs, but there are important differences. The laser source for photomixing delivers continous wave from two beams with slighly difference frequency which corresponds to the optical beat frequency(which is the desired THz frequency) unlike in TDS system where the laser source provides short pulses. The carrier density in the LTG-GaAs is modulated at the optical beat frequency, inducing a THz field in the antenna with Silicon hemispherical lens attached at the rear end of the dipole antenna couples the emitted

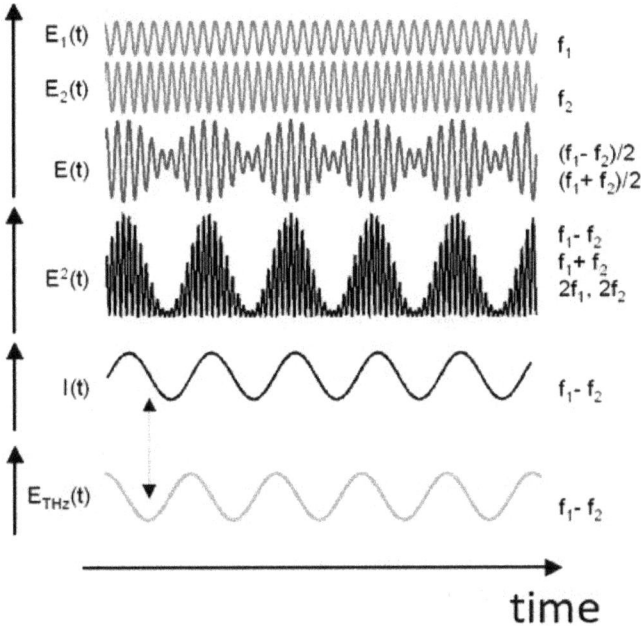

Figure 5.6: Proof of principle for THz photomixing of two signals with slightly different wavelengths for CW THz generation [90].

THz signal into free space [89]. In a photomixer the surface conductivity and charge carrier excitations are similar to those in photoconductive antenna. The number of photoexcited carriers $n(t)$ in the photomixer follows the rate equation [85].

$$\frac{dn(t)}{dt} + \frac{n}{\tau_{rec}} = \frac{\eta I_{exci}(t)}{E} \tag{5.9}$$

where η indicates the quantum efficiency, $I_{exci}(t)$ is the excitation pump intensity, τ_{rec} is the recombination time and $E = h\nu$ is the photon energy of the exciting beam and n is the carrier density. The exciting intensity on the surface of the semi-conductor antenna is as a result of the contribution of the two-laser beams. Figure 5.6 shows the basic principle of THz generation with photomixing. To demonstrate THz photomixing mechanism, consider two wavelengths with frequencies f_1 and f_2. They are superimposed on top of each other to generate an optical beat signal. Considering the electric fields of the two signals f_1 and f_2 are given by [89].

$$E_1(t) = E_1 sin(\omega_1 t) \tag{5.10}$$

and

$$E_2(t) = E_2 sin(\omega_2 t) \tag{5.11}$$

If the two exciting laser beams are linearly polarized and precisely overlapped, the resultant electric field will be a sum of the two electric fields given by

$$E(t) = E_1 sin(\omega_1 t) + E_2 sin(\omega_2 t) \tag{5.12}$$

As the photoconductor senses the intensity of the incident light, the instantaneous intensity of the two fields $I_{exci}(t)$ is related to the square of the total incidence electric field by [89].

$$I_{exci} \propto E^2(t) \tag{5.13}$$

Expanding equation 5.13 gives

$$I_{exci}(t) = E_1^2 sin^2(\omega_1 t) + E_2^2 sin^2(\omega_2 t) + E_1 E_2(cos(\omega_1 - \omega_2)t - cos(\omega_1+\omega_2)t) \tag{5.14}$$

Since the photoconductor converts the incident photon into photoexcited carriers the resultant signal is governed by the electric response of the photoconductor. The terms that vary at optical frequencies are ignored because the photoconductor cannot respond to rapid response with much faster timescale than its photocarrier lifetime [91]. The first and second terms of the equation have a time average value of $\frac{1}{2}E_1$ and $\frac{1}{2}E_2$ respectively resulting in DC terms while the term of $E_1 E_2 cos(\omega_1 + \omega_2)t$ contains optical frequencies. Hence, it will oscillate within a very short timescale i.e shorter than the carrier recombination time τ_{rec}, its contribution to $I_{exci}(t)$ is thus negligible since the electrons in the antenna follow the slowly oscillating frequency term [92]. The third term $E_1 E_2 cos(\omega_1 - \omega_2)t$ shows that the resulting term is the difference between the two incidence signals. The overall intensity response will be given by

$$I \approx E_1 E_2 cos(\omega_1 - \omega_2) \tag{5.15}$$

Eq 5.15 shows that the overall field frequency corresponds to the optical beat signal of the two signals.

The generated beat frequency is then focused on a biased photoconductive gap fabricated from a semiconductor material. Electrons and holes are generated by the optical beat frequency in the photoconducting gap. The electrons are accelerated and transits the valence band into the conduction band, if the photon energy of the beat frequency is large enough than the photoconductive gap energy. This process result to the flow of a photocurrent. The photocurrent is modulated at the optical beat frequency in the photoconducting gap generating a THz frequency, which is emited into free space by either antenna structure or silicon hemispherical lens implanted at the rear end of the photoconducting gap. The detected THz radiation frequency of the time varying signal, $\omega_1 - \omega_2$, can be varied by detuning the relative wavelength difference of the two incident pump lasers. The combined instanteneous power incident on the photoconductive antenna gap is described by [91, 93].

$$P(t) = P_o + 2\sqrt{\eta_m P_1 P_2} \left(cos(\omega_1 - \omega_2)t\right) \tag{5.16}$$

where η_m is the mixing efficiency which describes how best spatial overlap of the two modes is and ranges between 0 (no overlap) and 1(perfectly overlapped). The THz field $E_{THz}(t)$, emitted into free space is proportional to the generated photocurrent

$$E_{THz}(t) \approx \frac{dJ}{dt} \tag{5.17}$$

while its detected power $P_{THz}(t)$ at the receiving photoconducting antennna is given according to [91].

$$P_{THz}(t) = \frac{I_{DC}R}{(1 + (\omega_{THz}\tau)^2)(1 + (\omega_{THz}RC)^2)} \qquad (5.18)$$

and depends on the incident optical powers P_1 and P_2, the electrode bias voltage V_{bias}, the charge carrier lifetime τ, the antenna resistance R, capacitance C in the gap and the optical beat frequency $\omega_{THz} = \omega_1 - \omega_2$, the bias voltage is related to I_{DC}^2 by $I_{DC}^2 = V_{bias}^2 (P_1 + P_2)^2$.

Coherent detection can be accomplished with a similar photoconductive antenna using a fraction of the signal that is used for THz emission, with a time delay introduced such the signal to be detected and the probe signal reach the detector at same time. The detected photocurrent signal will be a convolution of the THz electric field with the laser modulated conductivity, which can lead to very high signal-to-noise ratios (SNR). In this case, the photoconductive antenna is not biased, but just the optically generated charge carriers recombine without the generation of current. However, in case an antenna structure is connected with the electrodes to detect the incoming THz signal, a voltage $U_{THz}(t)$, which is equivalent to the THz field strength will be generated at the THz frequency which is given by [93].

$$U_{THz}(t) \sim E_{THz}cos(\omega_{THz}t + \varphi) \qquad (5.19)$$

Similarly, the rate of generating the charge carriers is modulated with similar THz frequency and is given by

$$n(t) \sim P_L(t) = P_L(1 + cos(\omega_{THz}t) \qquad (5.20)$$

Conversely, the velocity of the carriers will be modulated by the detected THz radiations $v(t) \sim \mu U_{THz}$. The resultant photocurrent will be equivalent the product of carrier velocity and its concentration

$$I(t) \sim n(t)v(t) \sim cos(\omega_{THz}t)cos(\omega_{THz}t + \varphi) \qquad (5.21)$$

in which the photoconductive antenna acts as a homodyne mixer producing a dc component given as

$$< I > \sim < U_{THz}(t)P_L(t) > \sim E_{THz}P_Lcos\varphi \qquad (5.22)$$

The phase φ results from the optical delay between the optical signal and the THz signal before arriving at the receiving antenna for detection. Therefore, the photoconductor is a homodyne mixer capable of simultaneously detecting both the amplitude and the phase [93].

5.5 Properties of photoconductive materials

The carrier lifetime, τ_c, in transmitting as well as receiving photoconductive material is essential parameter for a wide spectral frequency response of the emitted THz signal. It is also important in determining the amount of power generated by the photoconductive antenna. At high frequencies the amount of the generated THz power is limited with the carrier lifetime of the photoconductive materials, therefore such carrier lifetime should have a short time scale inorder to realize high emission power [94]. The fundamental features for the photoconductive material suitable for photomixers include:

- a sub-picosecond carrier lifetime; which ensures ultrafast response of PC antenna for current modulation is achieved.

- huge dark break-down field; this will ensure large voltage is applied for the PC to radiate more and intense THz electric fields,

- large carrier mobility; for high efficient THz generation and

- a large photoconductor dark resistivity levels; to ensure that the DC electrical load is reduced [93], [94].

For 800 nm wavelength range the most commonly used semiconductor material is GaAs grown at low temperature.This beacuse it satisfies most of the above-mentioned requirements for photomixers. It is grown on a GaAs substrate which is sub-insulating at a low temperature of between 200–300°C and thermally annealed for 5-10 min at($\sim 600^o C$)after growth. Low temperature grown GaAs(LTG-GaAs) shows sub-picosecond carrier lifetime ($\tau < 1ps$), high-breakdown electric field ($E_B > 5 \times 10^5 V cm^{-1}$), high resistivity after annealing and large electron mobility ($> 200 cm^2 V^{-1} s^{-1}$) for huge photoconductive gain, making it attractive in building efficient photomixers [85, 89, 95, 96]. Despite large amount of work demonstrated so far with LTG-GaAs, its properties as well as the absolute temperature for growing the epitaxial layer, are challenging to control somewhat because there is lack of a reliable approach to monitor low temperature of $\sim 200^o C$ in high vacuum. Consequently, the properties of LTG-GaAs layers could change during the growth process even though the other growth features are unchanged. This hinders the optimization of the growth parameters and perfection of PCA devices made from these materials.

In that regard, ion implantation development in the recent past is promising approach for growing the substrate material for ultrafast PCAs as it yields good transition results for THz emission [85, 96]. Ion implantation is based on creation of defects by destroying the gallium Arsenide (GaAs) lattice structure. The fabrication mechanism for ion implantation is based on defect engineering in which GaAs is grown at low temperatures resulting to ions not settling in the right way, for example Arsenic (As) ions are located in Ga sites and Ga vacancies. This create what is referred point defects. These defects provide large recombination cross-section areas for the charge carriers pairs (electrons-holes pairs), eventually reducing the carrier lifetime [84]. Subsequently, the grown LTG-GaAs is thermally annealed to increase its resistivity which eventually will prompt high voltage to applied giving rise to high power photocurrent [76].

The use of ion-implanted GaAs has several benefits compared to LTG-GaAs. For example, though quite challenging to control the growth temperature of the properties of the LTG GaAs technology in the MBE chamber the ion dosage implanted material properties can be precisely and reproducibly tailor-made. Furthermore, ion implantation offers the possibility to vary precisely the features of the semiconductor substrate by changing the appropriate implantation and annealing conditions [76, 96]. The electrodes are patterned with an interdigitated finger structure on its top surface to enhance, homogenize the electric field distribution and maintain the active area the same. The output power can be scaled up by increasing the applied field and the capacitance C, however at the expense of limited high-frequency performance [76]. In this study we employed a structure similar to the one shown in Figure 5.7.

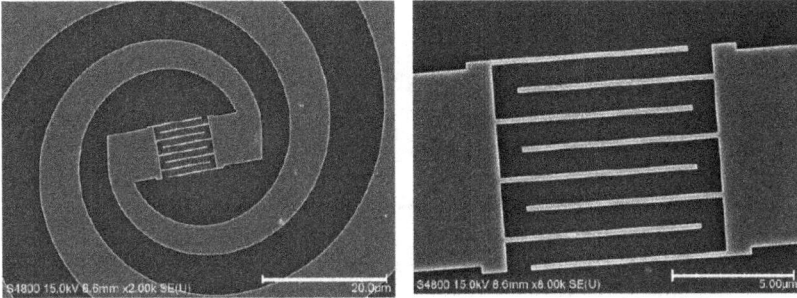

Figure 5.7: Picture of scanning electron microscope (SEM) images of self-complimentary log spiral antenna with the interdigitated finger structure [76].

5.6 Two-color laser concepts for terahertz photomixing

Different two-color laser concepts have been developed for THz wave emission through the process of photomixing. These concepts include the use of two independent laser sources with slightly detuned frequencies [89], or two-color laser systems, either with an external cavity control element or monolithic integration of two lasers for emission of two modes simultenuously from same cavity [97]. Earlier works by Brown et al, demonstrated THz photomixing with two independent ultra-short pulse lasers such as Ti:Sapphire lasers [98]. Since ultra-short pulses are rather complex, expensive, and bulky, the generation of THz radiation with such a system turns out to be expensive and challenging. Alternative approach to employ compact and cost-effective semiconductor lasers was proposed to circumvent the challenges encountered by Brown et al and to push THz technology out of the lab into real world application. Shortly after Brown's work, McIntosh et al [99] demonstrated the generation of THz frequency of up to 5 THz using two independent Distributed Bragg reflector (DBR) diode lasers. Figure 5.8 shows a photomixing setup employing two independent laser diodes as excitation source for photomixers. Most THz photomixing systems in the market utilize two independent semiconductor lasers as excitation sources for the photomixers. Much concentrated effort is needed to align and precisely superimpose the two beams to obtain a good optical beat signal for efficient current generation in the photoconductive antenna with two independent laser sources [89, 97].

One laser frequency is tunable while the other laser frequency is fixed. The two modes from the lasers are then combined in beam combiner before their optical beat

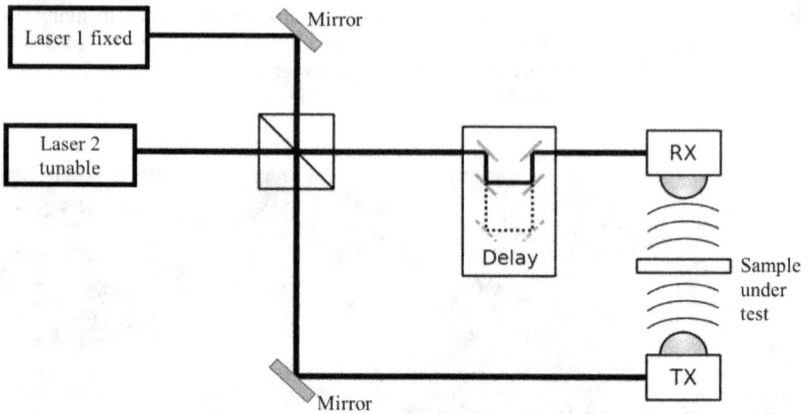

Figure 5.8: Schematic showing THz photomixing with two independent laser sources.

is split into two for THz generation in the transmitting antenna (TX) and detection with receiving photoconductive antennas (RX) [98] Such systems will require several optics for the tight alignment of the beams which eventually will result to a large, more complex, and costly system. Furthermore, the generated THz radiations spectral purity in the photomixers depends entirely on the quality of the beat signal, which also is dependent on how stable the two exciting modes are. Stable THz signal emission demands that the exciting lasers be stable. This can be achieved by stabilizing each pump laser source independently and overlapping the two modes from the two independent lasers precisely to obtain a good optical beating signals. All this requires concentrated effort in order to obtain a more stable optical beat signal. Such optical alignment and stabilization of the optical beat frequency challenges can be easily overcome by employing dual mode lasers, which are designed to emit two frequencies simultaneously either from a single or combined laser cavity [100]. Similarly, dual wavelengths propagating in the same cavity experience common frequency fluctuations. These fluctuations compensate each other at the difference frequency as the two modes propagates the same cavity thereby resulting to more stable signal than the frequency of individual laser mode [2]. Stable optical beat frequency will result to a stable THz radiation generation. The concept with two color emission either with external cavity operation or monolithic two-color laser operation will be discussed in the following subsection.

5.6.1 Two color diode lasers based on external cavity configuration

To achieve stability and wavelength tunability of semiconductor laser diodes, diode lasers are usually arranged in an external cavity configurations. The generated THz radiation spectral purity via photomixing is highly dependent on the purity of the optical beat, which is governed by the frequency stability of the pump lasers [89]. External cavity dual mode concepts based on frequency-selectable filters or grating mirrors as excitation source for CW THz systems have been reported by various groups [91, 101]. Gu et al, demonstrated a tunable coherent CW-THz radiation source by photomixing with a dual-wavelength laser diode in an external cavity configuration using V-shaped double-stripe end mirror as a means of selecting the two modes simultaneously [101]. Later on, Kleine-Ostmann et al, used a similar concept of two-color external cavity source to demonstrate CW-THz imaging with high signal to noise ratio [102].

This concept is a modified Littman cavity configuration with an additional lens in the cavity. The beam of the semiconductor diode laser is collimated and focused onto a diffraction grating. The first order diffracted beam is focused onto the end mirror which provides the optical feedback for the different desired modes while the zeroth order mode is the output. A V-shaped structured end mirror is used to select two wavelengths for an optical beat signal emission [103] as shown in Figure 5.9. However, the structured end mirror permitted multiple modes to oscillate in the external cavity limiting the spectral purity of the optical beat signal [104]. Instead of a V-shaped end mirror, Park et al, used a beam splitter to split the first order beam into two. With the use of two end mirrors, he managed to demonstrate emission of two frequencies simultaneously in what was described as tunable double external cavity configuration (2λ-ECSL) with an improved spectral purity of the selected modes [91].

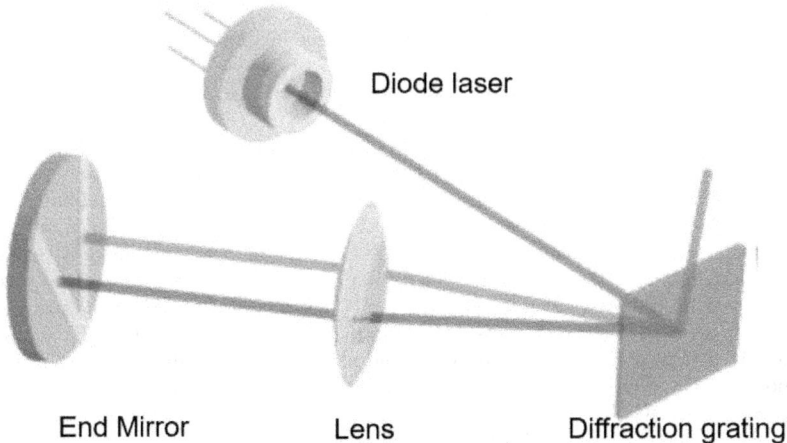

Figure 5.9: Schematic of a Fourier transform external cavity diode laser having a V-shaped strip end mirror for different mode selection for THz generation [103].

Another interesting concept is based on Littrow configuration in which two diffraction grating are arranged in Littrow configuration and can be operated simultaneously giving dual wavelength emission with pure spectral selectivity [104]. Fredrich et al, analyzed different two-color concepts and observed that external cavity configurations are more flexible and permit the adjustment of the optical beat frequencies over broad frequency range. Nevertheless, depending on their geometrical realization, two color external cavity lasers exhibit varied characteristics with respect to spectral linewidth, maximum and minimum optical beat frequency, and signal to amplified spontaneous emission (ASE) ratio; more details on this can be found in ref [31]. Despite these systems exhibiting high output emission powers and broad frequency tuning range, they are quite large as they have many mechanical moving parts, complexity in beam alignment, and require high investment costs. Therefore, their inherent configuration has hindered ECLs from being deployed to develop small size and low-cost CW THz emitters [105]. Monolithically integrated DBR laser emitting two modes simultaneously provides a stabilized optical beat frequency for THz photomixing and are quite small (just a single chip) hence more promising than ECLs.

5.6.2 Monolithically integrated two-color lasers

The monolithically integrated two-color laser is a CW laser, particularly designed to emit two wavelengths simulteneously with their frequencies having a small offset. The emitted frequencies can be employed to generate a single-frequency THz signal via photomixing of the optical beat signal in a photoconductive medium. The optical beat, from dual mode coherent laser, determines the purity, stability, and noise characteristics of the emitted THz wave. When two modes are emitted from a single cavity and travel in same waveguide the noise in the generated CW THz wave is minimal. This is because the frequency fluctuations emanating from thermal or mechanical changes cancel each other resulting to a stable difference frequency [89, 105]. The optical gain of a two-color semiconductor laser is broad hence it can generate several optical beat signals even up to several THz. In addition, the semiconductor laser is quite small, and it is easier to operate in comparison to other type of lasers [89].

Several groups have demonstrated two-color lasers with DFB and DBR laser structures being the most studied, for example Iio et al [106] demonstrated a laser device which had a periodic-phase-shift grating in the DBR section enabling the selection of two wavelength laser emission simultaneously. The operation mode of the device was determined with the period of the grating while the mode spacing was determined by the interval at which the phase shifts [89]. Such structure was employed by Hidaka et al [107] and Gu et al [108] to generate THz radiation via photomixing. In this case, however, the spectral tuning of the generated THz signal in this structure got lost as a result of the fixed frequency difference of the two modes [105]. Another example was a structure developed by Uemukai et al, they developed a AlGaAs dual mode DBR laser on a single substrate and managed to generate tunable THz radiation in the range 0.8-1.2 THz via photomixing using LTG-GaAs photomixer with stable optical beat signal, however the optical power for each mode was observed fluctuating [109].

Monolithically integrated dual modes have become key for THz generation due to the following advantages:

- They provide stable optical beat signal. The oscillations of two modes in similar waveguide superimpose each other, resulting to the cancellation of any fluctuations. Subsequently, the yielded beat signal is more stable in comparison to the signal from a distinct laser system [109]

- Easy overlap of the modes, the two modes can be easily and precisely overlapped [89, 110].

In the recent years, a lot of research development have been witnessed focused at possibility to improve two-color monolithically integrated semiconductor laser diode used as excitation sources for photomixing in terms of stability, expanding the tunable frequency range as well as emission powers have been reported [111–114]. Nevertheless, achieving stable, tunable, and high-power emission properties at once on a single chip still are open questions which need to be investigated further to make these devices available for application.

In this thesis, we demonstrate CW THz sources based on monolithically integrated two-color diode lasers. A first source was a compact CW THz system with stable emission based on two-color DBR laser. The developed CW THz source was based on photomixing of a two-color signals using fiber coupled photoconductive antennas. The generated 300 GHz signal was used to investigate its potential application in THz non-destructive testing applications for moisture measurement on leaf and different papers [115, 116]. The second terahertz source, was based on a new DBR monolithically intergrated diode laser. The latter is electrically tunable and is designed with capability to adjust the optical beat frequency of the emitted wavelengths. Optical beat tuning was accomplished with micro resistor heater implemented on top of the DBR section. It changes the local temperature of the DBR segment via carrier injection thereby shifting the emission wavelength as a result of the changes in the refractive index. We demonstrated terahertz wave generation in the frequency range of 100 GHz-300 GHz via photomixing. Furthermore, we demonstrated the potential use of the implemented THz spectrometer system in non-contact measurement for the thickness determination of a sample of PE block and the spectroscopic transmission characteristics of a THz filter.

5.7 Device description and experimental setup

In this section, the devices used in this study i.e two-color DBR lasers will be discussed and the experimental setup for THz generation will be highlighted. Two types of devices will be discussed.

5.7.1 A two-color DBR diode laser

In the first case, we employed a two-color monolithically integrated DBR semiconductor diode laser what I refer here a two-color DBR laser to distinguish it from the second device. The two-color DBR laser is a single chip laser emitting two modes simulteneously at the center wavelength of 785 nm and has Y-waveguide which ensures that the two modes have a common output. This DBR device was employed as excitation source for the photomixing. The structure of the DBR laser has four sections: two ridge waveguide (RW)-sections having a gain section and a DBR section, a Y-shaped waveguide which joins the two RW sections together to obtain a single common output section. This device was 3000 μm long with an active region of 2.2 μm . The two ridge waveguide branches where realized based on a sine-generated curve to form Y-shaped bends. 10th order surface DBR gratings with slightly different grating periods where used to define the wavelength of the two modes with a separation of 0.6 nm. A similar structure to the one used in this study is described in ref [28]. However, the structure employed in this investigation was mounted p-side down with all the four contacts having a common contact for injection current hence they where operated simultaneously with a single injection current. For the experiment, the laser diode was mounted on a copper heatsink with a thermo electric cooler attached at the bottom of the mount to minimize any thermal flunctuations. 90 mW optical power was measured directly in front of the laser without any optical amplification with an injection current of 200 mA . The diode signal was collimated and coupled into a polarization compensating single mode fiber to compensate for the different beam divergences of the two colors, enhance the spectral alignment and precisely overlap the modes on top of each other for an enhanced optical beating signal. The schematic structure of the two-color DBR LD used in this study is as shown in Figure 5.10 [29] and was fabricated at Ferdinand Braun Institut (FBH) Berlin.

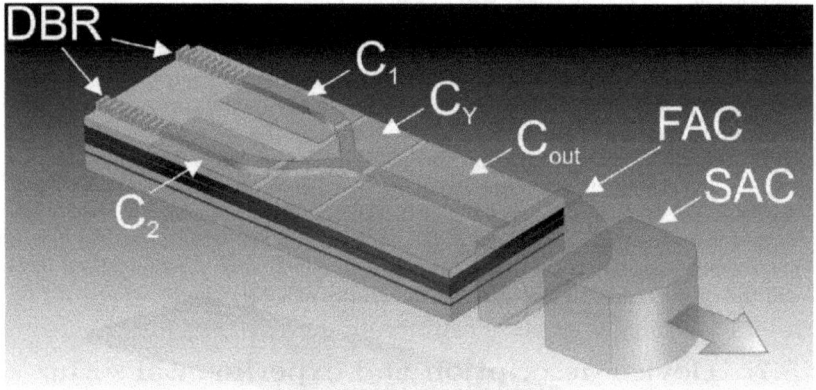

Figure 5.10: Schematic illustration of the Y-shaped two color DBR semiconductor laser. The rear end consists of two DBR mirrors, and C_1, C_2, C_Y , C_{out} , are the electrode contacts for current injection [29].

5.7.2 An electrically tunable two-color DBR diode laser

In the second case, we employed an electrically tunable monolithically integrated two-color distributed Bragg reflector semiconductor laser what have refered a two-color eDBR laser, emitting also at 785 nm center wavelength fabricated also at FBH by metal organic vapor phase epitaxy (MOVPE) [117]. Two DBR gratings are implemented at the end of the laser chip for wavelength selection, two ridge waveguides branches guide the laser beams and a Y-shaped waveguide section was implemented to realize a common output of the two modes. The two different modes are determined with 10th order surface DBR grating with small difference periods with a spectral separation of 0.6 nm between the two main modes. A resistor micro heater is implemented on top of DBR section which allows the independent adjustment of wavelengths of each DBR section as it only induces the temperature change of the local area below it. This is achieved via carrier injection thereby changing the refractive index of the DBR section, in the process shifting the operation wavelength of the laser diode.

The two-color eDBR laser has five separate electrical contacts, enabling individual current injection to the two branches: two contacts to the ridge waveguide(RW)sections (I_R, I_L), a contact for the microheater (I_{HC}), a contact to the Y-branch (I_Y), and a single contact for the output section (I_{out}) as shown in Figure 5.11. The output contact enables a current injection to the output waveguide and helps to maintain power balancing of the two lasing modes for efficient mode beating. The independent current injection to each section enables faster spectral tuning than the approach we employed in our previous work [116] in which wavelength adjustment was attained by changing the temperature of both the laser chip and laser submount with the use of a thermo-electric controller concurrently. In addition to that benefit, wavelength tuning with micro resistor heaters ensures that the optical beat signal can be continuously adjusted without mode hoping and it preserves the spectral linewidth of the lasing modes [118]. The device is 3000 μm long in length. The section lengths are 500 μm, 1350 μm,200 μm and 900 μm for the DBR grating, RW section, Y-branch and common section, respectively. The width of the RW is 5 μm and the two DBR-branches have a separation distance of 80 μm which provides electrical isolation between the two branches. The distinct electrical contacts allow an independent current injection to each DBR section enabling individual control of the wavelength emission. The DBR laser was mounted p-side up on a conduction cooled package (CCP) which has a 25 mm by 25 mm footprint as shown in Figure 5.12. The CCP was then mounted into an aluminum heat sink which was thermally stabilized with thermo electric cooler (Peltier element) and $10k\Omega$ thermistor was used for completing the thermoelectric loop. For the optical characterization of the diode we implemented a setup as shown in Figure 5.13. The emitted optical signal through the output section was collimated with an aspheric lens with a focal length of 6.24 mm and a numerical aperture NA = 0.42 (Thorlabs, A110TM-B). Two plain mirrors direct the collimated beam into a fiber collimation package with focal length of 8.0 mm (Thorlabs, F240FC-780) for coupling the signal into a single mode fiber. Similar devices with same concept design are described in more details in ref [117, 119].

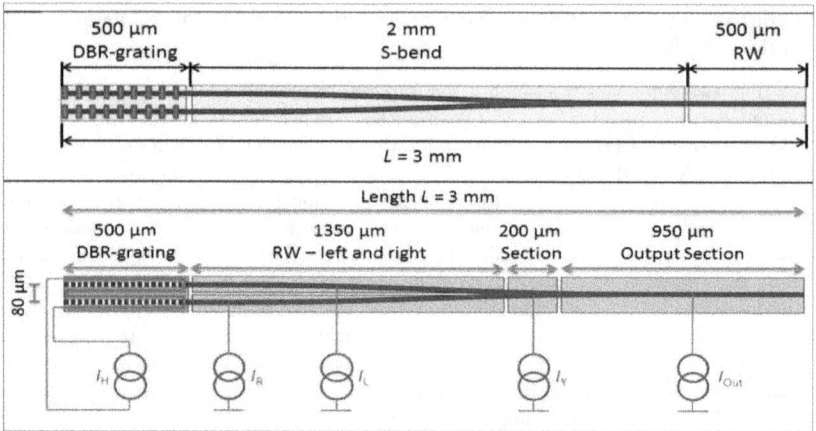

Figure 5.11: The structure of the monolithically integrated two-color laser with separate electrical contacts labelled I_H, I_R, I_L, I_Y, and I_{out} for independent current injection to each section. [117].

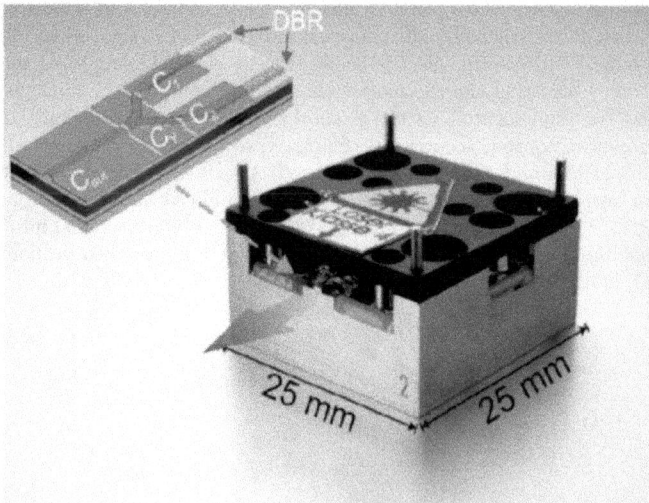

Figure 5.12: The structure of the monolithically integrated two-color laser implemented on a conduction cooled package. [119].

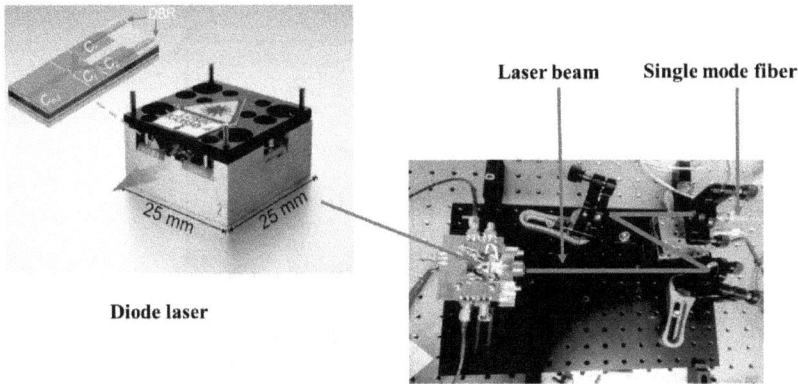

Figure 5.13: Picture showing the diode laser implemented in a characterization setup

5.7.3 Experimental description for CW THz generation and detection

In this study, we employed an homodyne experimental setup to generate and detect CW THz radiation as shown in schematic Figure 5.14. The emission and detection of the CW THz signal was accomplished with two identical logarithmic spiral photoconducting antennas made of ion-implanted Gallium Arsenide (GaAs) with interdigitate finger electrodes in the photoconductive gap. Silicon hyper-hemispherical lenses are attached at rear end of the photoconductive antenna to radiate the generated signal into free space. The photomixers have eight 0.2 μm interdigitated fingers separated by 1 μm electrode gaps with an active area of 9 x 9 μm^2. The active region is fabricated on 1 μm thick LT-GaAs substrate with Ti:Au metallization made in the middle of a three-turn logarithmic spiral antenna [80]. Both transmitting and receiving antennas are fiber coupled with polarization maintaining fibers since they are sensitive to the exciting laser beam. The optical beam from the two color DBR laser was split off into two signals using an optical beam splitter. One beam was focused into the photoconducting transmitting antenna (PCA_t) for THz generation. The second optical beam was directed into the photoconducting receiving antenna (PCA_r) through a motorized delay stage for THz detection of the generated signal. The optical path length of the PCA_r was varied in order inorder that both the detecting optical signal and the generated THz signal are syncronized at the receiver at the same time. Both the phase and amplitude of the generated THz radiation are determined. In the PCA_t, the optical beat from the DBR laser modulates the photoconductive antenna and varying carrier densities are generated exciting the electrons to transit from valence band into conduction band. A photocurrent is generated at the emitting antenna when it is supplied with bias voltage, having its frequency equivalent to the difference frequency of the exciting two laser modes and was detected with the log spiral antenna at the detecting antenna as a THz electric fields. With an average of 5 mW optical power at the antennas, a 30μA photocurrent was detected at the PCA_t while for PCA_r photocurrent was 21μA measured with a Keithley 2636A system source meter. A sinusoidal bias volt-

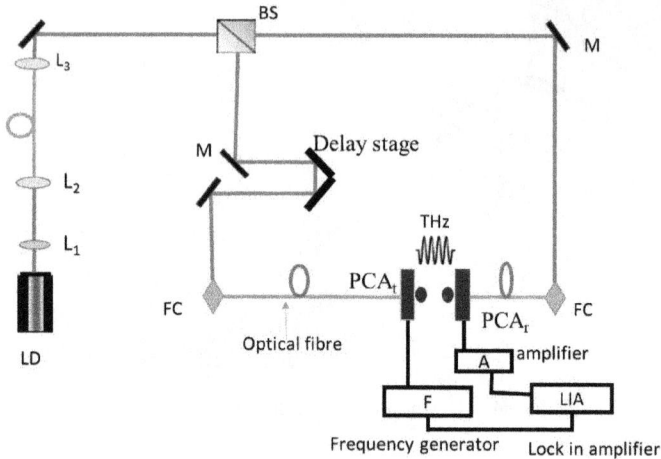

Figure 5.14: Schematic illustration of an homodyne setup for THz generation and detection with fiber coupled photoconductive antennas. LD: The two color distributed Bragg reflector laser, BS: Beam splitter, (PCA_t): transmitting photoconductive antenna, (PCA_r): Receiving photoconductive antenna.

age of ± 4V, was applied at the transmitting antenna and modulated with 2.18 kHz for THz emission.An hemispherical silicon lens incorporated in the photoconductive antenna radiates the generated THz signal into free space and focuses it into the receiving antenna for detection. A transimpedance amplifier was employed to amplify the generated photocurrent before its response was measured by a lock-in amplifier with an integration time constant of 300 ms.

5.8   Results and discussion

In this section, optical emission characteristics and THz emission characteristics for both devices are discussed. The potential application of the assembled THz sources will also be presented in this section.

5.8.1   Measurements with two-color DBR laser

5.8.2   Optical characteristics of the two-color DBR laser

Before we employed the two-color DBR laser for THz generation we evaluated its optical characterists, with measurements being performed at $15^o C$ room temperature and laser operated in a continuous mode. The optical characteristics were obtained with Yokogawa AQ6370B optical spectrum analyzer(OSA) with a spectral resolution of 0.02 nm, equivalent to 9.6 GHz frequency resolution by varying current and temperature of the diode. The P-I measurements where obtained by varying the injection current and determining the corresponding power with power meter PI100(Thorlabs). Figure 5.15 (a) shows the output power verses the injected current obtained directly in front of the diode. The P-I characteristics curve shows that the device can generate up to 90 mW of output power when the laser is excited with 200 mA injection current. The spectrum measurements were made at 200 mA injection current with 10 mW of average power coupled into a single mode fiber for analysis. Figure 5.15 (b) shows the obtained optical spectrum. For these settings, the spectrum exhibits two color emission with left mode having a wavelength of 785 nm and the right mode a wavelength of 785.6 nm with a spectral difference of 0.6 nm, which is proportional to 296 GHz difference frequency.

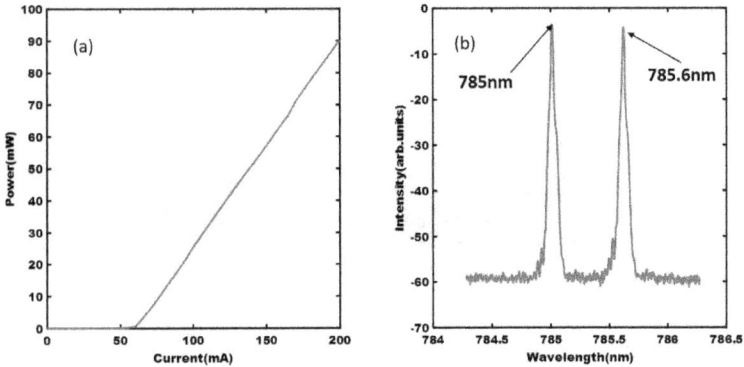

Figure 5.15: (a) VI characteristic curve. (b) Optical spectrum from two-color DBR Laser diode.

To investigate how stable the system is, measurements of the difference frequency in dependence of current where carried out. First we examined the intensity distribution in the two modes by varying the injection current. Figure 5.16 illustrates a contour plot showing the intensity map of the two modes. The threshold current at the left mode is 57 mA while the right mode is 70 mA. Mode jumps can be clearly observed at around 60 mA, 70 mA, 90 mA, 140 mA for the left mode while in the right mode the mode jumps are witnessed at 88 mA, 114 mA and 128 mA. Sudden mode transition between different laser modes at the mentioned current can be observed. The mode hopping noticed in this structure is undesirable for stable optical generation which is a requirement for stable THz emission. It could contribute to intensity fluctuations and affect the power balancing to the two main modes desired for efficient optical beating at the antenna.

5.8.3   Stability characterization

Due to the mode hopping mentioned in section 5.8.2 further investigation were done to find the potential stable operation regimes for operation of this device for THz measurements. In this regard optical beat tuning measurements was performed in dependence of the injection current and temperature variation [115, 116]. Figure 5.17(a) illustrates the different spectra obtained at an injection current of 200 mA by adjusting the temperature of the laser chip and the entire diode sub-mount

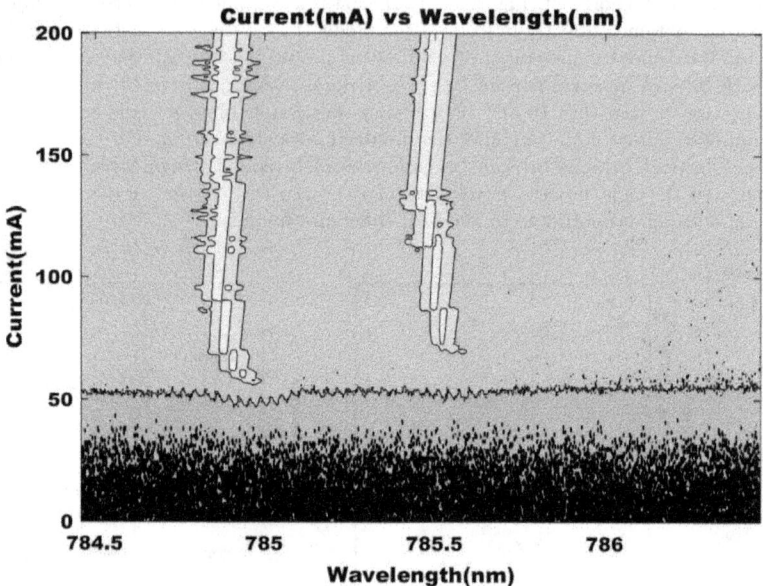

Figure 5.16: Contour plot showing intensity distribution in the two optical modes.

from 15^oC to 21^oC. The results show spectral adjustment with constant spectral separation of 0.6 nm of the two modes. This give an optical beat signal equivalent to $\Delta v \sim 296$ GHz. In Figure 5.17(b) it can be observed that at currents below 180 mA, and within the 15^oC to 21^oC temperature range, only three discrete difference frequencies are emitted. We attribute this to mode hopping in the different arms of the Y-structure. However, above 180 mA injection current, the device exhibited a relatively stable difference frequency for all the measured temperature range. At various temperature settings, the difference frequency can be adjusted from 280 GHz to 320 GHz. This corresponds to a bandwidth range of 40 GHz.

From Figure 5.17 (b) by varying the temperature and current settings it can be observed that the difference frequency appears centered around 300 GHz. The obtained characteristics for two color DBR laser shows that could be suitable to generate CW THz radiation in the spectral bandwidth from 280 GHz to 320 GHz. This bandwidth was obtained by changing the temperature of the sub-mount plus that of the diode laser concurrently. For fine tuning of the optical beat signal, a new structure which allows independed current injection to the different section of the DBR laser which ensures individual mode tuning was investigated and the results are presented in the proceeding subsection. The new device spectral mode positions are fine tuned with the changes in the refractive index of the semiconductor material in the DBR section which are adjusted via the carrier density. Moreover, the separate current injection to each section of the diode permits an active stabilization of the optical beat signal enabling one to obtain a clean THz signal with even lower phase noise [115].

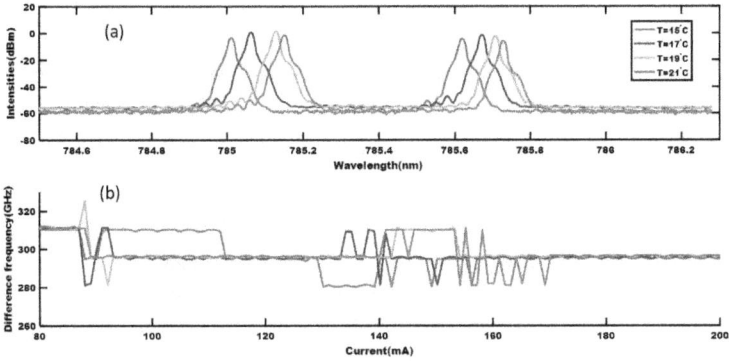

Figure 5.17: (a) Spectra of optical beat tuning at various temperature settings between 15^oC to 21^oC (b) Different frequency measurement with various temperature tuning for CW THz generation.

5.8.4   CW THz generation and detection with two-color DBR laser

As mentioned before a coherent homodyne setup was employed to generate and detect the CW THz radiation. Before the THz measurement were done, we checked that both modes are emitted from our device simultaneously. The intensity of the two modes was monitored with optical fiber spectrometer (Avaspec 2048). This was necessary to ensure an excellent beating signal was obtained at the photoconductive antenna. The combined output of the two colors was sent to the receiving antennas, a photocurrent was generated due to the applied bias and depended on the introduced delay time. Figure 5.18 (a) shows the obtained photocurrent measurement, which exhibits the generated THz fields with its frequency corresponds to the optical beat frequency of the two colors from the DBR laser diode. Figure5.18 (b) shows the obtained frequency from the Fast Fourier transform (FFT) of the photocurrent. The FFT shows the generated THz frequency as a single frequency at 0.3 THz frequency. This frequency was quite in agreement with the optical beat frequency of the two wavelengths emitted by our DBR laser diode structure which was obtained by the OSA measurement.

The linewidth of the main peak in Figure 5.18 (b) was evaluated depended on the accuracy of our THz measurement. The available measurement was obtained with a time delay of 25 ps, we estimate a frequency resolution of $\Delta f \sim 40$ GHz which coincides with the width of the peak by chance. Thus, the THz linewidth could fall below 40 GHz, if a longer delay time will be employed and this could be consistent with the optical spectra in Figure 5.15 (b). The generated THz power was approximated from the photocurrent of the photoconductive antenna and its value was estimated to be in lower μW range, this was so since during the measurements, there was no calibrated THz detector for power measurement which was available. To emit more THz power, it is expected that the exciting optical power of the laser to be high, the device investigated here demonstrated 90 mW of optical power with 200 mA injection current, similar devices delivering around 200 mW of output optical power have been reported [28]. Moreover, in the subsection 5.8.6 we analyze a new DBR device with higher emission power which permits separate currents to be sent into the individual segments. The new device enable us to vary the difference frequency of the two colors more precisely. The available difference frequencies discussed in this section are discrete because they are determined by mode jumps.

5.8.5 Potential application

We explore the potential use of our system in non-destructive moisture measurement with experiments performed on pieces of printing papers with different moisture content, and on a leaf subjected to drought induced stress for a few days before the plant is watered to see the plant recovery from the drought induced stresses. The aim was to demonstrate the potential of this system for real deployment application in the field.

Paper industry

The papers have great potential in several sectors of the economy contributing to various fields ranging from education, communication, historical data storage, etc. The drying process, is a vital process of paper production and is a major factor which determines on the final quality of the paper. The uniform distribution of moisture during the drying process contributes to the curling, shrinkages, strength etc. properties of the paper. Non-uniform moisture distribution profiles could cause various unwanted problems ranging from printer misfeeding, blackening of printed area, curling, increased tearing and cockling of paper [122]. Monitoring the moisture content of the papers during the repackaging process to ensure only dry pieces of paper are sorted for storage or packing is therefore important. In this regard, we

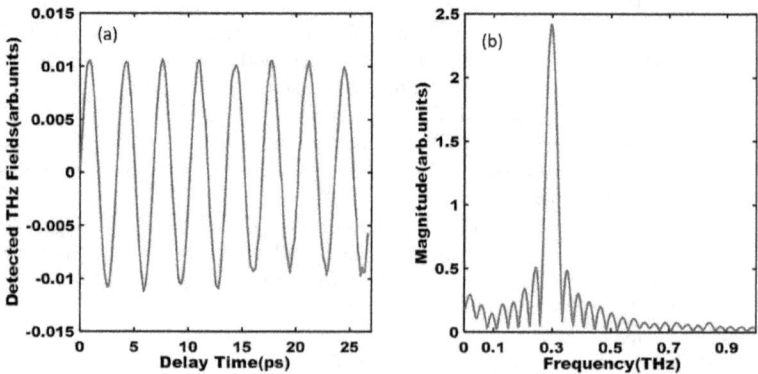

Figure 5.18: (a) The generated photocurrent in the transmitting photoconducting antenna. It corresponds to the THz electric field detected with receiving photoconducting antenna. (b) The obtained THz frequency by finding the Fast Fourier Transform of generated photocurrent with the main peak centered at 300 GHz.

demonstrate the use of our 0.3 THz source in simple online monitoring of the drying process of two pieces of printing papers having different moisture distribution on their surface. One piece of paper has a drop of water on the center where THz radiation is focused before THz measurement is taken, while the second piece of paper was fully saturated in water before placed in the THz radiation path for THz measurements. In Figure 5.19 the detected THz signal amplitude changes are plotted against drying time. An increase in THz transmission amplitudes over time as the pieces of paper dry up is observed. With a sheet of paper with drop of water showing higher amplitude levels compared to the saturated one since it takes a shorter period to dry. Upon the paper drying a constant THz amplitude level is attained. It is presumed the leveling of the amplitude to infer reduced moisture content hence paper could be dry and ready for storage.

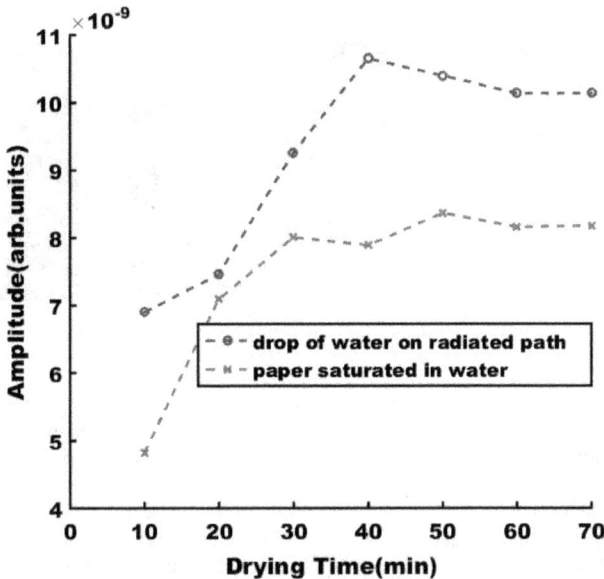

Figure 5.19: THz signal changes detected during drying process of two pieces of paper with different moisture concentration.

Plant water status monitoring

As the ecosystem is experiencing unprecedent weather changes, water which is an essential component for plants life is becoming scarce hence the need for prudent and intelligent planning of its use in watering plants. An approach to predict the plant water status in order to enable for timely planning for irrigation will be beneficial to plant biologists as well as greenhouse farmers. Since plants play a significant role of stabilizing the ecosystem as well as food supply their timely salvage from wilting is necessary. Monitoring the plant water status with non-destructive testing approach

Figure 5.20: Experimental setup with the leaf inserted on the path of THz radiation.

can provide useful information for timely plant watering. In this investigation, the area in the leaf used for the THz measurements remained the same throughout the entire experiment. This was necessary to avoid two problems: (i) any variation which may be related to putative location variations at area of measurement because of taking measurements at different points due to moving the plant into and out of the THz radiation path [123]. (ii) The likely increase in disturbances, which may influence the plant's stress response [124]. Figure 5.20 illustrates a setup with a leaf under test. The dry leaf tissue may have some influence on the transmitted signal, however THz absorption in water may result to noticeable attenuation of the signal that can be used monitor the leaf's water status directly. Since THz radiation is highly absorbed by water, the detected THz intensity passing through the leaf could be correlated to the amount of the water in the plant [123]. Thus, the transmittance can be used to monitor the plant water status. The measurement was done after the watering was suspended for few days to induce drought stresses to the plant and measurement taken over several days without watering the plant. The THz transmission strongly increases indicating a significant decline in water status of the plant as shown in Figure 5.21. The increase of the water content after re-watering of the drought stressed plant occurs much rapidly, due to the quick water uptake.This results in the transmittance decreases within a few hours. Moisture being a good absorber of THz radiation, there is a rapid decrease in transmission amplitudes as the plant recovers from drought same trend is observed in regions where rewatering was done. This is clearly illustrated with the expanded section of the plot on Figure 5.21. Accordingly, we observed this process by measuring the THz transmission of the leaf in steps of 1-hour interval. It is interesting to observe that the water uptake by the plant could be monitored efficiently as shown in expanded region of Figure 5.21. Studies made using more complex THz-TDS systems [124–126] and with terahertz quasi time domain spectroscopy(THz QTDS) [127] show similar trends. We have shown here a more compact and cost-effective source with comparable results. The major advantage of our setup is compactness. It is possible to repackage all the components of our setup into a single small box which will make it easy to transfer it into the field.

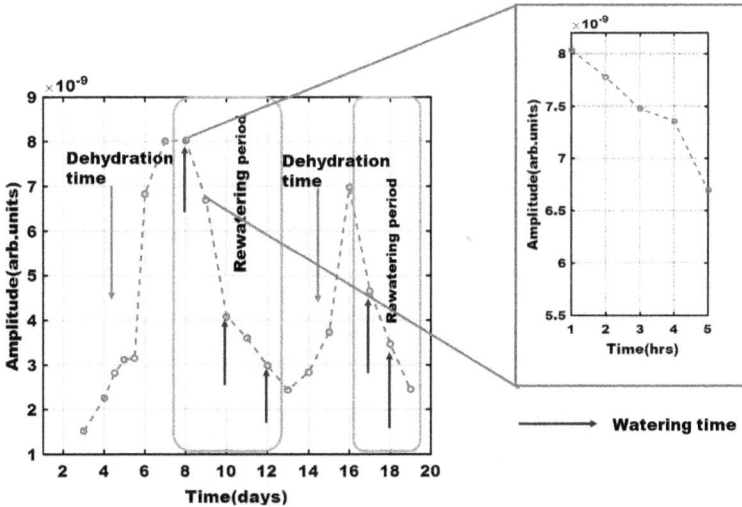

Figure 5.21: THz transmission amplitude of a leaf experiencing dehydration and rehydration. The green region represents the rewatering period of the plant. With the black arrow indicating morning and evening watering periods on same day.

5.8.6 Measurements with an electrically tunable two-color DBR laser

Figure 5.22 shows a VI characteristics curve of the eDBR diode laser device with the fiber coupled powers versus the injection current from 0 to 200 mA for the individual two-color laser modes. When the output current is on, there is a slight increase in the emitted powers with a reduction of the threshold current to 20 mA in comparison to threshold of 80 mA when the output section is not supplied with current. Figures 5.23 and 5.24 shows the emission wavelength tuning spectrum for each laser when operated independently. From this measurement, the tuning range of DBR laser 1 is from 784.4 nm to 785.6 nm with current injection to micro heaters adjusted between 0 to 350 mA. From DBR laser 1 wavelength span of 1.2 nm is obtained. While, DBR laser 2 under similar injection current can be tuned from

785.0 nm to 785.4 nm which gives a wavelength difference of 0.4 nm. DBR laser 1 exhibited a broad wavelength tuning range in comparison to DBR laser 1, this was possible as the micro resistor heater was implemented closer to its grating than to DBR laser 2.

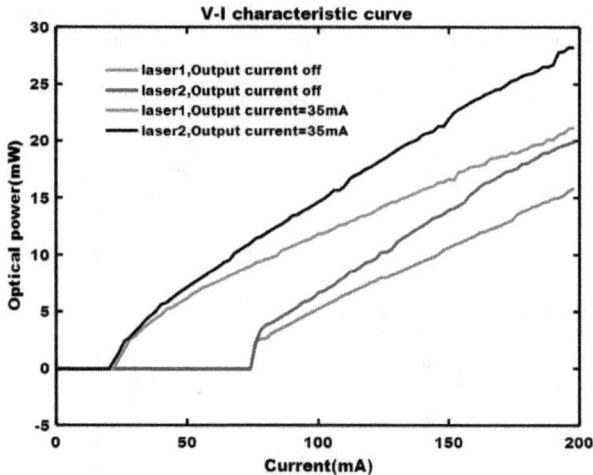

Figure 5.22: VI characteristic curve for the individual DBR laser when the output is on and off.

5.8.7 Optical characteristics with micro resistor heater current off and output current off

Figure 5.25 shows the spectra measurement which were obtained with one laser on while the other was switched off, under similar injection current of 60 mA with the heater current off and the output current off ($I_{HC}=I_{out}=0$), with the heat sink temperature set at 20°C. DBR laser 1 emits at 784.5 nm with DBR laser 2 emitting at 785.1 nm. The spectra measurement when both the lasers are operated simultaneously is depicted in Figure 5.25(b) clearly showing the stable emission at 784.5 nm and 785.1 nm. The spectral separation of this emission gives an optical beat signal of $\Delta v \sim 290$ GHz frequency. The mode spectral distance can be changed when a current is supplied to the micro resistor heater and varied from 0 to 350 mA. Furthermore, a side mode suppression ratio (SMSR) greater than 45 dB was obtained. These lasing spectra were measured with an OSA (Yokogawa AQ6370B) with a spectral resolution of 0.02 nm. When both the lasers are lasing simulteneously, the intensity observed in both the wavelengths are almost equal. This is important since for THz photomixing, it is desirable that the powers of the two wavelengths are equal for efficient mode beating at the antenna.

Figure 5.26 shows measurements made at various selected injection currents to evaluate the influence of higher injection currents on the emitted laser modes. There is an increase in output power level plus the SMSR of more than 45 dB maintained,

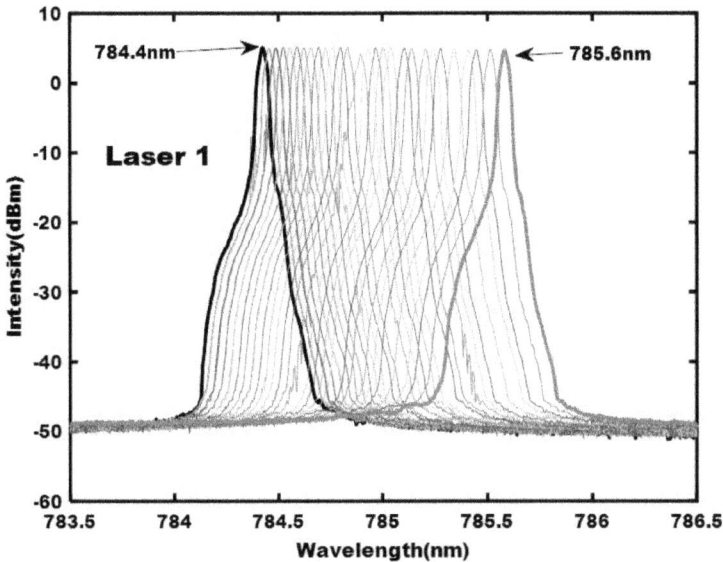

Figure 5.23: Optical spectrum showing the tuning of DBR laser 1 from 784.4 nm to 785.6 nm with microheater current settings of 0 to 350 mA.

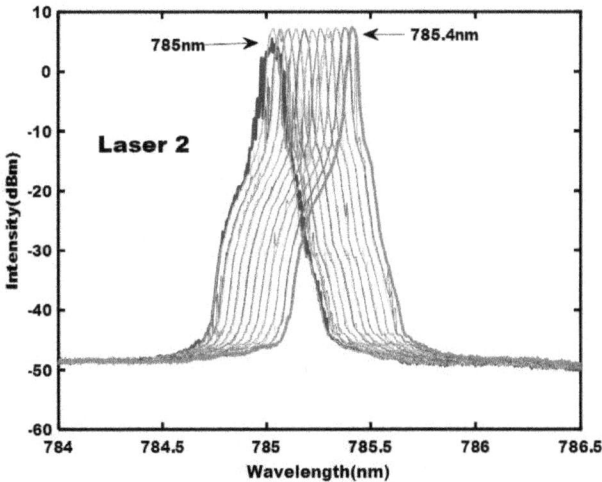

Figure 5.24: Optical spectrum showing the tuning of DBR laser 2 from 785.0 nm to 785.4 nm with microheater current settings of 0 to 350 mA.

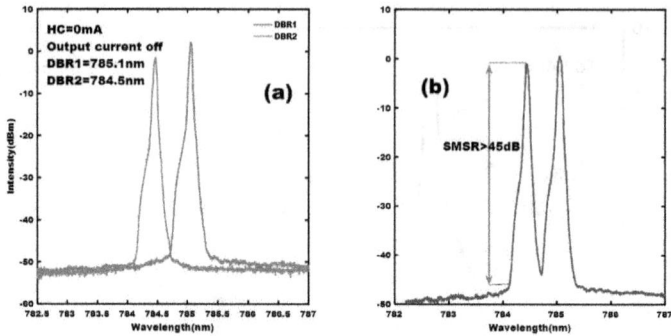

Figure 5.25: (a) Optical spectrum for both lasers when one laser is on while the other is off (b)Dual mode optical spectrum when both the lasers are operated simultaneously.

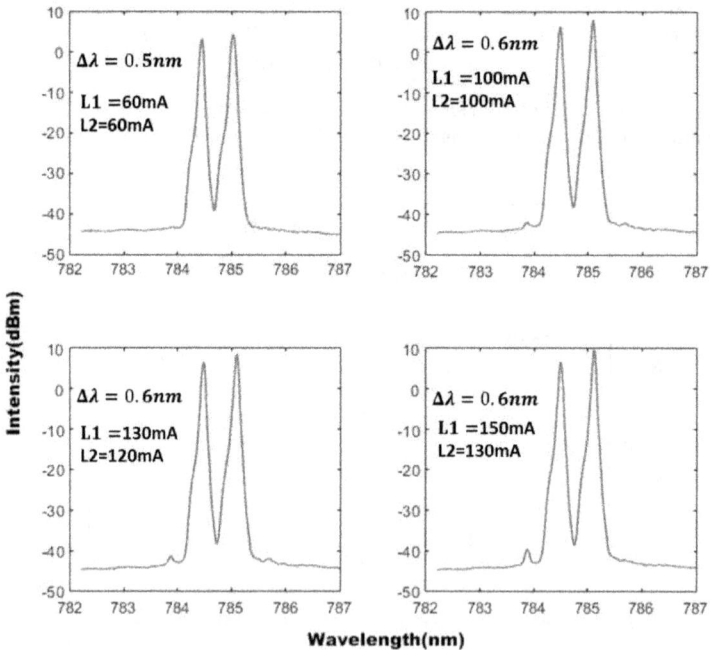

Figure 5.26: Optical spectra illustrating the response of the modes with increasing injection currents.

when the current of both the two modes are increased. However, an interesting feature with this structure was the appearance of side bands besides the main modes at higher injection currents in the optical spectrum. These phenomena appeared because the two colors undergoes four wave mixing (FWM) when they are emitted simultaneously. The strong appearance of the sidebands clearly demonstrates good spatial overlap of the two modes, polarization, and with minimum phase noise [64, 81]. This clearly demonstrates a strong indication that this device would provide an efficient optical beating frequency at the photoconductive antenna for THz emission.

In Figure 5.27 a single optical spectrum in terms of FWM sideband was analyzed for the measurement taken when the laser was operated with higher injection currents of $L_1 = 165$ mA, $L_2 = 160$ mA, with both output current and the heater currents off. The observable sideband occurs with 0.6 nm spectral difference from the main mode on both sides. The lower FWM side band occurs at 783.9 nm while the upper side band occurs at 785.7 nm. The modulation of the charge carrier optical properties such as optical gain with the optical beating signal leads to ultrafast nonlinear changes in the carrier distribution thereby generating the sidebands on the sides of the main optical modes [64]. Despite the presence of FWM side bands that occur under the influence of optical beat modulation condition, the two modes of the laser are strongly harmonized to generate efficient and strong difference frequency signal as they exhibit higher SMSR [128].

5.8.8 Optical characteristics with micro resistor heater current on

The spectral seperation of the two wavelengths are adjusted by varying temperature of the local area of the DBR section through changing the refractive index thereby shifting the emission wavelength. Here, the spectral behavior was investigated with respect to the current injection to the resistor heater implemented on top of the DBR region. Figure 5.28 shows the continuous optical spectral emission tuning characteristics for both two colors DBR1 and DBR 2 with the adjustment of some selected heater currents from 100 mA to 210 mA. Waterfall Figure 5.29 depicts the tuning of the two modes with current injection to the resistor heater. The modes at higher heater currents are seen to be converging to each other which eventually results to mode overlap. Injection current to both the lasers was $L_1 = L_2 = 60$ mA, while the heater currents were varied from 0 mA to 250 mA.

We investigate further how broad we can tune our diode laser by increasing the heater current to 350 mA. Figure 5.30 illustrates the wavelength adjustment of both modes with increasing heater injection current. The resistor heater currents were set in steps of 5 mA. For the measurement these heater currents where electrically scanned from 0 mA to 350 mA with measurements taking approximately 6 minutes. The spectral distances can be adjusted even much faster by increasing the step size of the heater currents. The local temperature of the DBR region was increased instantly resulting to the shifting of the wavelengths thereby adjusting the spectral separation between the two emission modes. We observe both wavelengths being shifted at the same time towards longer wavelengths. While tuning the spectral emission it was observed that DBR1 emission mode shift towards DBR2 emission mode until both modes lase at same wavelength at higher heater current of around 250 mA. The injection current between 0 and 250 mA provided tunable optical

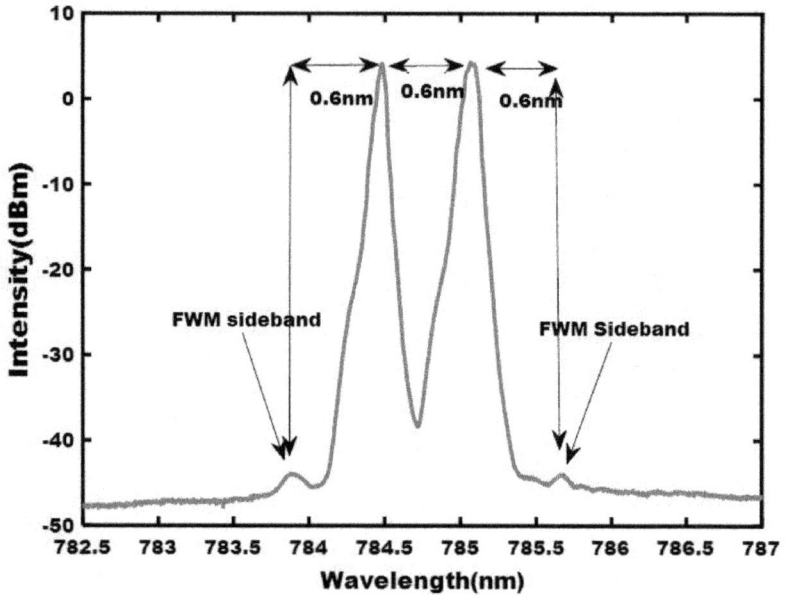

Figure 5.27: Optical spectrum with Four wave mixing side bands on both sides of the main lasing modes. Heater current=0 mA, DBR1 current=165 mA, DBR2 current=160 mA, output current=0 mA main wavelength modes are laser1=784.5 nm and laser2=785.1 nm.

beat frequencies was only limited due to the mode overlap observed at 250 mA injection heater current. To investigate how broad the optical beat could be tunable, measurement for the optical beat frequency against injected heater current was done as shown in Figure 5.31. It clearly shows that the optical beating frequencies could be tunable between 100 GHz to around 280 GHz. At higher heater injection currents in the region between 250 mA-350 mA, the modes overlap resulting to the difference frequency of zero hence no optical beating is obtained. This limits the tuning bandwidth range of our optical beat frequency and hence the THz signal to be generated.

5.8.9 Optical characteristics with optical spectrum analyzer having higher spectral resolution

The charaterization of the optical spectra with an optical spectrum analyzer(OSA) with higher spectral resolution reveals that the diode operates in multimode. In this case, the OSA Yokogawa (AQ6370D) having 0.01 nm spectral resolution was employed to analyze the optical signal. Figure 5.32 shows the obtained optical spectrum signal showing some sidemodes alongside the two main modes. This spectral measurement reveals that our device is a running in more longitudinal modes next

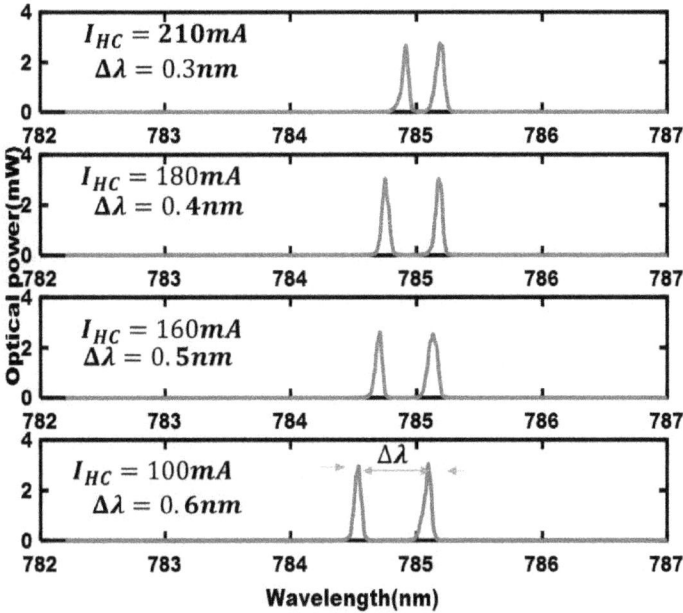

Figure 5.28: Wavelength tuning characteristics of electrically tunable DBR laser obtained as a result of detuning of the spectral distances with increasing resistor heater currents.

to the main modes. The longitudinal modes present at the two main modes will contribute in the generation of multiple THz signal frequencies in the photomixing process. This is because each competing optical mode contributes to the overall optical beating signal which results to multiple THz frequency generation in the antenna.

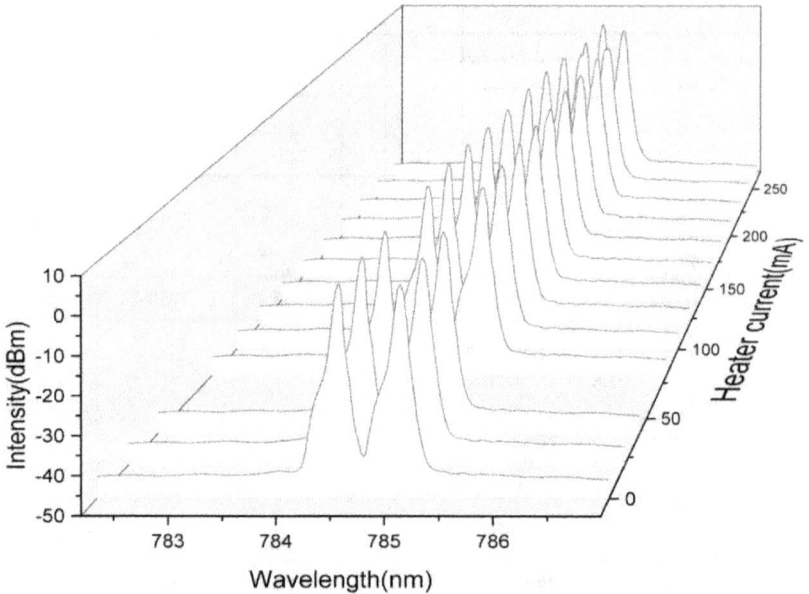

Figure 5.29: Waterfall picture of the two wavelengths tunable while both the DBR lasers are operating simultaneously. Clearly showing the decreasing spectral distances at higher injection heater currents.

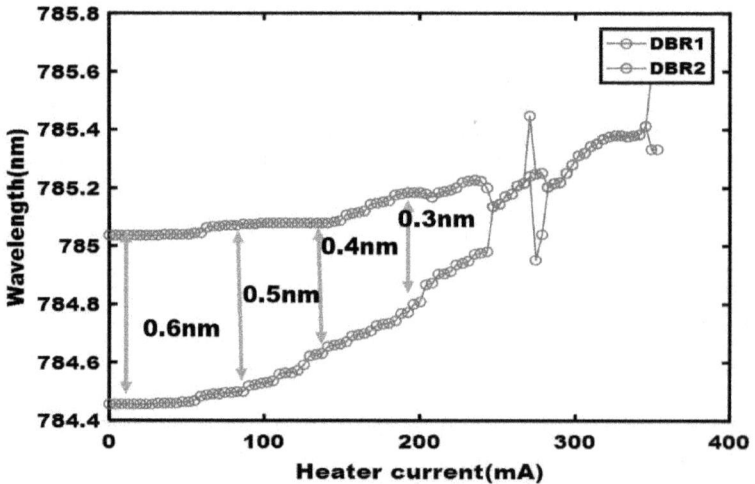

Figure 5.30: Characteristics of the two wavelengths tunability with both the DBR lasers operating simultaneously with increasing heater currents. Clearly showing mode overlap at around 250 mA injection heater currents.

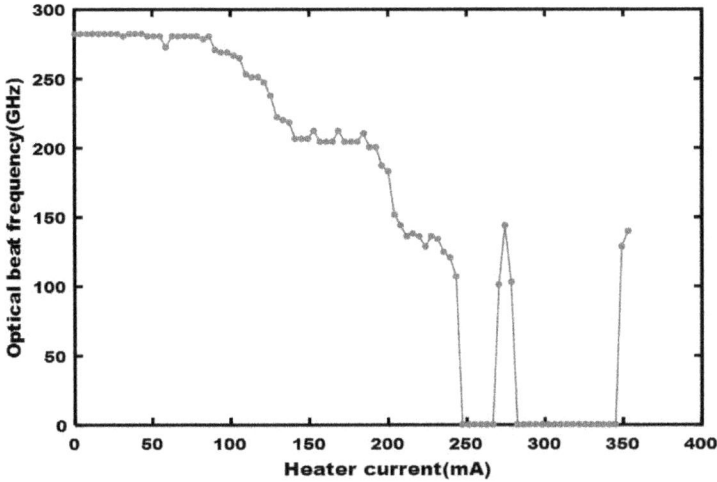

Figure 5.31: Tunability of the optical beat frequency with increasing resistor heater currents. The optical beat could be tunable from 100 GHz to 290 GHz.

5.8.10 Terahertz emission and detection with two-color eDBR laser

Single frequency THz signal generation

We demonstrated the potential use of the eDBR laser as excitation source for single frequency THz generation. THz photomixing of the two-color eDBR laser signal generated the CW THz waves which was confirmed in a homodyne detection scheme similar to one described in 5.7.3. However, in this investigation, THz lenses were incorporated on the THz path in order to ensure much of THz signal power was delivered to the receiver for detection and the transmission of THz signal through the sample for spectrocopic characterization of the sample properties was enhanced. In this way the losses of THz signal due to scattering at the sample which was analyzed were minimized. The new THz path with Teflon and TPX lenses is shown in Figure 5.33. For the demonstration of a single frequency THz signal generation, the delay stage was scanned from zero position toward the positive side. The generated THz photocurrent signal and the corresponding THz frequency spectrum are shown in Figure 5.34. The photocurrents recorded vary periodically over the introduced delay time and corresponds to the generated THz electric field in the emitter antenna. 0.3 THz frequency was obtained by finding the FFT of the generated

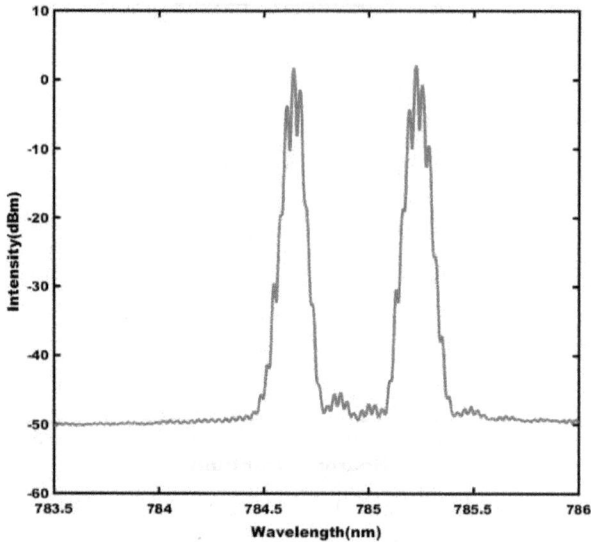

Figure 5.32: Optical spectrum signal taken with higher spectral resolution of 0.01 nm with Yokogawa AQ6370D optical spectrum analyzer.

photocurrent signal, which is in agreement with the optical beat frequency signal of the two main modes. We demonstrate the generation of various THz frequencies by detuning the spectral separation of the two modes by changing the resistor heater currents. Figure 5.35 shows some selected THz photocurrent measurements and the corresponding THz frequencies are shown in Figure 5.36 obtained by FFT of the generated photocurrents. In this set of measurements, the THz signal was varied from 0.182 THz to 0.3 THz by adjustment of the resistor heater current which varies the spectral separation between the two modes. These measurements were obtained by changing the heater current from 0 mA to 200 mA.

Multi-frequency THz signal generation

To investigate the THz signal linewidth, a longer delay stage was used. Experiments were repeated similar to onces perfomed to generate single THz frequency signal. However, in this investigation the delay stage was scanned a distance of 12.2 cm

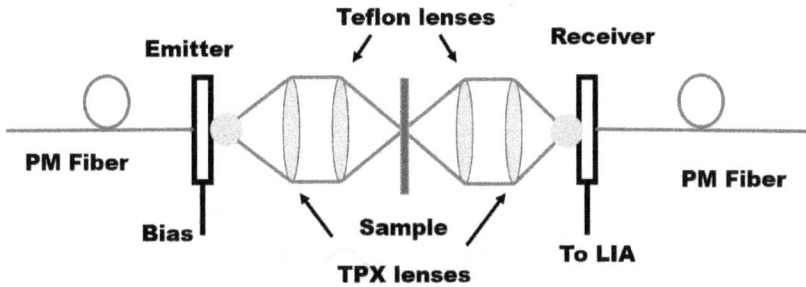

Figure 5.33: THz path with THz lenses incorperated to ehnance coupling of signal to the receiving antenna

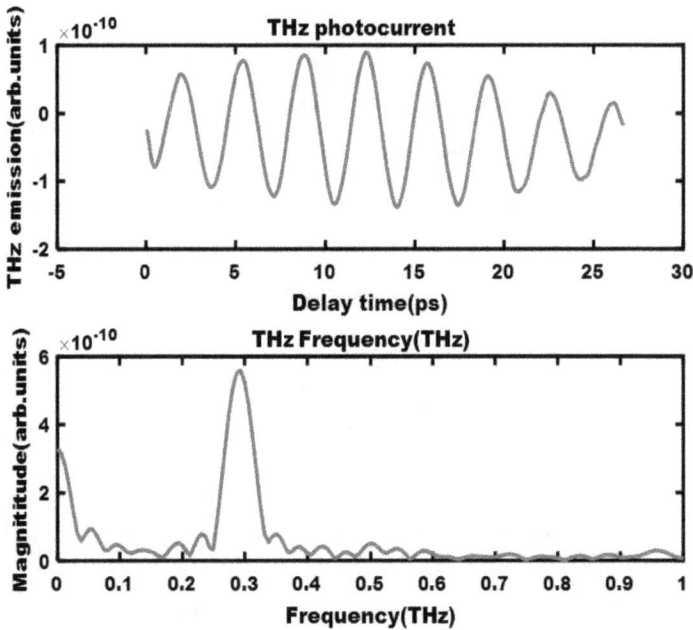

Figure 5.34: Generated THz photocurrents and the corresponding THz frequency at 0.3 THz frequency when the laser diode is operated at $L_1 = L_2 = 80$ mA with heater current=0 mA and output currents=0 mA.

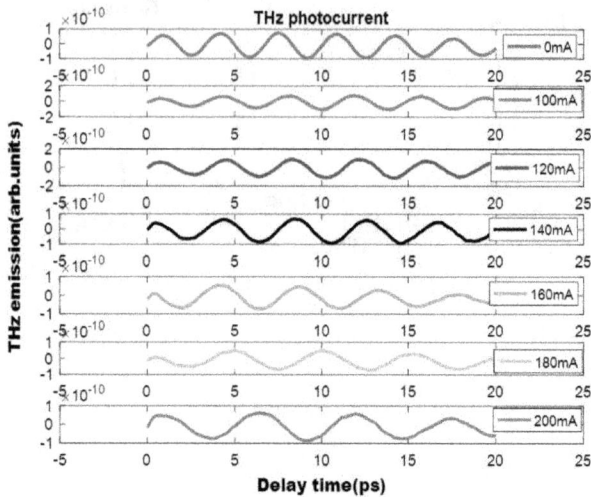

Figure 5.35: Generated THz photocurrents due to the detuning of the lasing modes with the variation of resistor heater currents when laser diode is operated at $L_1 = L_2 = 60$ mA with output current=35 mA and some selected heater currents varied from 0 mA to 200 mA.

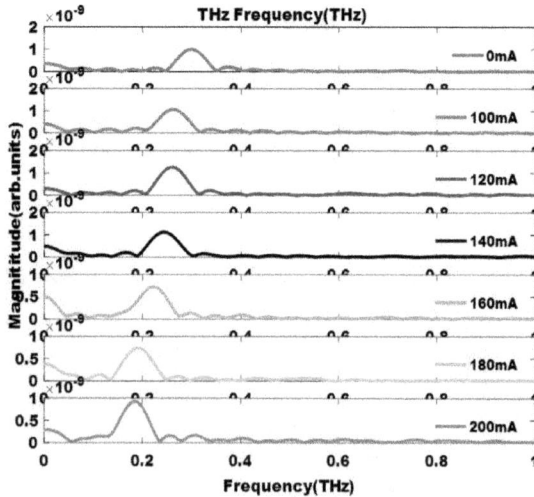

Figure 5.36: The corresponding THz frequency to Figure 5.35 showing tunable frequencies in the range 0.182 THz to 0.3 THz.

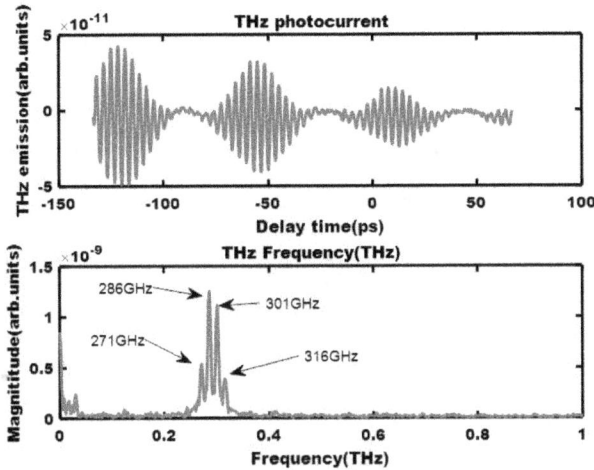

Figure 5.37: Photocurrent harmonics with the repetition of Signal amplitude at long delay time when the delay stage is scanned 12.2 cm. When the injection current of $L_1 = L_2 = 80$ mA. With the corresponding FFT shown with mult-frequencies centered around 0.3 THz.

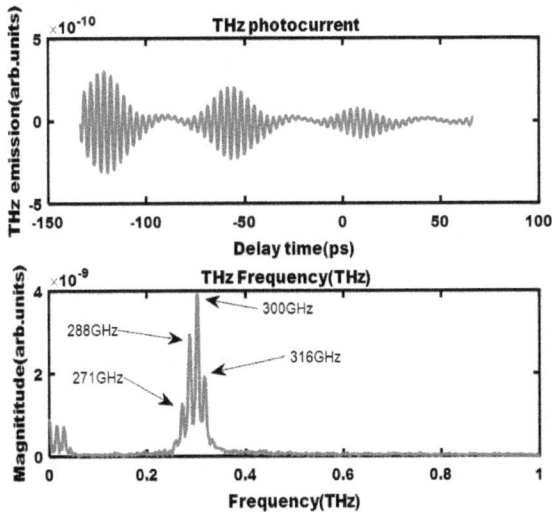

Figure 5.38: Photocurrent harmonics when the injection current of diode is increased $L_1 = L_2 = 120$ mA. The corresponding FFT shows frequency centered around 0.3 THz

starting from a negative position with the aim of enhancing the spectral resolution of the generated THz signal. We hoped to obtain a clean THz signal with high spectral resolution. But on the contrary, multiple THz frequency signal were generated. The multiple modes in the optical spectrum contributed to the addition frequencies generated. Three photocurrents spectral amplitudes harmonics are noticed in the measurement shown in Figure 5.37 with their amplitudes appearing to increase towards the negative direction. From these photocurrents multiple THz frequency centred around 0.3 THz are obtained by transforming the photocurrent signal into frequency domain as shown in Figure 5.37. The measurement was performed when the laser was operated at $L_1 = L_2 = 80$ mA with the output currents and heater currents switched off. The corresponding FFT of the photocurrent scan shows distinct frequencies generated at 0.271 THz, 0.288 THz, 0.300 THz and 0.316 THz. A comparison measurement was perfomed to investigate whether the generated photocurrent harmonics are obtained at higher injection currents. Similar photocurrent harmonics amplitudes are obtained at higher injection currents as shown in Figure 5.38. The generated multi-frequency THz signal contains frequencies with same values as the case for lower injection currents. The measurement was carried out with the laser operated at $L_1 = L_2 = 120$ mA injection currents, with the resistor heater current and the output current not in operation.

The measurements were repeated for generation of THz radiation at the $L_1 = L_2 = 120$ mA injection currents laser operation with a further scan of the delay stage towards a much more negative starting position. In this experiment, the measurement reveals three full maximum photocurrent amplitude spectra with the amplitude spectrum with much intensity centered around -150 ps to -100 ps as shown in Figure 5.39. The delay stage was scanned a distance of 16.4 cm. Four frequencies centered around 300 GHz are obtained as shown in Figure 5.40, similar to the two cases above.

The generation of the several photocurrent amplitudes was due to the multiple modes in the optical beat which contributed to the photomixing process in the antenna. The number of optical modes available in the optical beat signal for photomixing contributes to the number of THz frequencies generated at the antenna [130]. The multi-frequency THz signals generated is centered around 300 GHz which corresponds to the optical beat frequency value of the two main modes with spectral distance of 0.6 nm. This is due to the fact that the device investigated here was designed to emit two modes simultaneously at a spectral separation of 0.6 nm but due to unexpected developments it emitted multiple frequencies. The generated multi-frequency signals exhibit an equidistant spectral mode spacing of 15 GHz as shown in the inset of Figure 5.40, with their amplitudes appearing to reduce sideways from the central mode.

5.8.11   Potential applications of developed THz system

In this section, we discuss some examples where the developed THz system could gain application. In this regard, we investigated the potential use of the developed system in THz spectroscopic applications such as characterization of THz filter and in THz metrology in thickness measurement of polyethylene block sample using multi-frequency THz signal generated by the system by electrically tuning the diode laser.

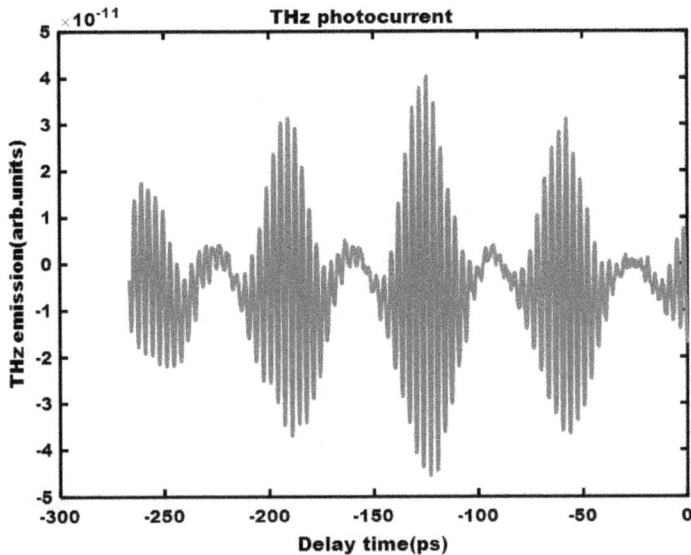

Figure 5.39: Long scan of the delay stage towards the negative side shows repetitions of several THz photocurrent harmonics spectrum amplitudes with the highest amplitude depicted between -150 to -100 ps.

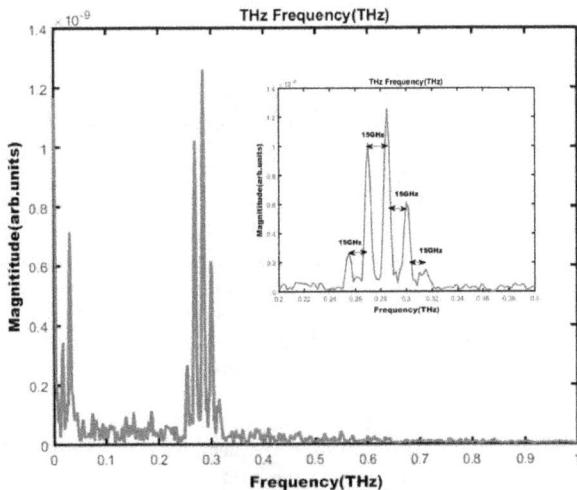

Figure 5.40: FFT of the photocurrent signal in Figure 5.39 showing generated THz frequency centered around 0.3 THz.

THz spectroscopy

The transmission characteristics of a THz bandpass filter (BPF250GHz-24 Tydex) were measured with our multi-frequency CW THz source. The bandpass filter was manufactured from a thin nickel foil material which was perforated with a cross-absent shaped pattern on its surface which act as the frequency selective surface. The filter has a resonant frequency centered around 250 GHz. The cross apertures are periodically patterned with the periodicity of 795 μm and the spacing between adjacent crosses is 175 μm. Both arms of the cross have same equidistance length of 620 μm. Figure 5.41 shows a microscope image of a segment of the analyzed THz filter. The transmission measurements were done by introducing the filter on the path of the THz signal, in between the transmitting and receiving antennas. THz photocurrent was measured with and without the filter on the path of the THz beam. The generated THz photocurrent signal measurements and its corresponding frequency signal measurements are shown in Figure 5.42. The THz signal without the filter is the reference measurement while the measurement with the filter (in green) is the sample measurement. The THz signal was attenuated when the filter was inserted onto the path of the THz signal. The photocurrent signal was Fourier transformed to obtain the frequency measurement as shown in Figure 5.42 (below). The obtained frequency measurements provide the reference electric field amplitudes as E_{ref} and the electric field amplitude due to the filter as E_{filter}. The transmittance of the filter (T_{filter}) was obtained by calculating the ratio of absolute values of the electric field's amplitudes between the filter and reference measurement given by

$$T_{filter} = \frac{|E_{filter}|}{|E_{ref}|} \qquad (5.23)$$

The transmittance characteristics of the band pass filter with our CW THz source are shown in Figure 5.43. The obtained data set indicates the peak transmittance at 252.8 GHz with transmission percentage of 93 % being the highest transmission recorded. The measurement was obtained by tuning the THz frequencies between 156.9 GHz to 315.3 GHz through the adjustments of the micro-resistor heater current of the diode from 0 to 200 mA.

A comparison study was done with a more complex THz time domain spectroscopy system (Tera OSCAT from MenloSytems) based on optical sampling by cavity tuning (OSCAT) technique. Figure 5.44 shows the reference measurement without the filter on the THz path. Figure 5.44 (a) shows the recorded lock in amplifier (LIA) signal showing terahertz transient without the filter and a signal obtained with a Butterworth filter with cutoff frequency of 400 GHz, while Figure 5.44(b) illustrates the Fast Fourier transform of the THz pulse signal in Figure 5.44 (a) with the amplitude spectrum was calculated using rectangular window and compared with another calculated using hanning window. The measurement provides the frequency spectrum of the THz signal without the filter. The obtained spectrum clearly shows the water absorption lines at 557.4 GHz and 752.6 GHz. Figure 5.45 shows the measurement with the filter inserted on the path of the THz signal. Figure 5.45 (a) shows the lock in amplifier signal showing terahertz transient obtained with the filter on the THz path and butterworth filter signal which cutoff frequencies above 400GHz. Figure 5.45 (b) shows the corresponding FFT of the THz pulse giving the amplitude frequency spectrum of the THz signal with the filter, the spectrum was calculated using rectangular window and compared with

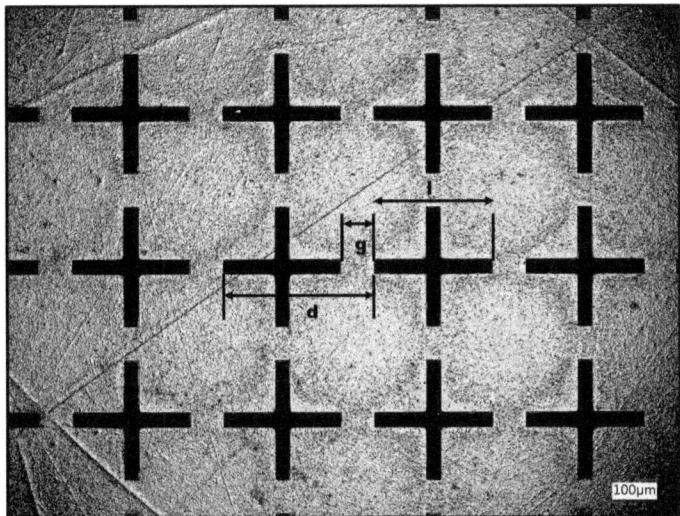

Figure 5.41: Microscope image of a segment of the 250 GHz THz Band pass filter. The periodicity d of the cross-shaped pattern is 795 μm and the gap distance g between the crosses is 175 μm. Both arms of the cross have same equidistance length l of 620 μm. Image and measurement were obtained with VHX Digital Microscope VHX-5000 with magnification of x100 μm.

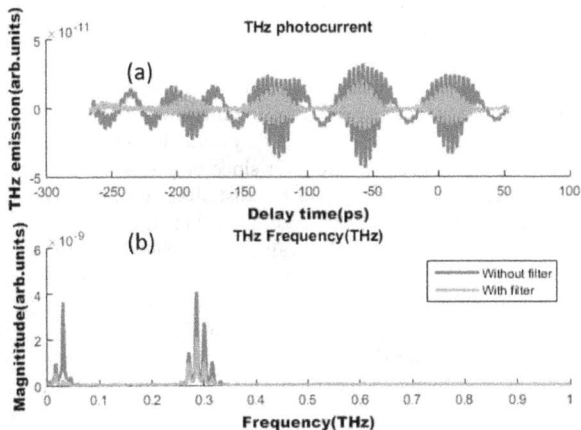

Figure 5.42: CW THz photocurrent signal measurements with and without the filter and the corresponding frequency measurement below.

the hanning window measurement. It clearly shows the peak amplitude centered around 250 GHz. From the frequency spectrums Figure 5.44 (b) and Figure 5.45

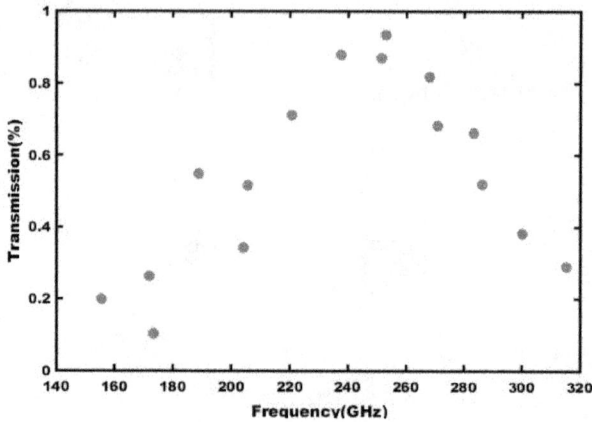

Figure 5.43: Transmission characteristics of the 250 GHz THz filter at different CW THz frequencies ranging from 156.9 GHz to 315.3 GHz.

(b), the transmittance characteristics of the filter was calculated by determining the ratio of the amplitudes between the reference measurement and filter measurement. Figure 5.46 show the obtained transmittance of the filter. This plot clearly shows the peak transmittance around 250 GHz with cutoff frequency around 150 GHz and passband frequency of 350 GHz with the OSCAT system.

In general, Figure 5.47 summarizes the transmission characteristics of the THz filter with the measurements done with our CW THz system, with the Tydex data and OSCAT system measurement. The obtained measurement shows the cutoff frequency is around 150 GHz with a peak of the transmittance for both the three systems centered around 250 GHz resonance frequency and a passband frequency of around 350 GHz for both Tydex and OSCAT system. For our system we could not estimate the passband frequency of the filter since our tunable range of our system was limited to 315 GHz, but the trend of the distribution of our data set measurement points to the passband obtained with the other systems. Our experimental data set is in close agreement with the transmission characteristics measurements provided in the data sheet by Tydex as well as with the measurements obtained with the OSCAT system. By inserting this filter into a THz radiation path, one can narrow a broadband THz signal bandwidth in the range 150 GHz to 350 GHz with peak transmittance centered at around a resonance frequency of 250 GHz. We believe our system would be an important tool for spectroscopic studies in the range 100 GHz to 315 GHz.

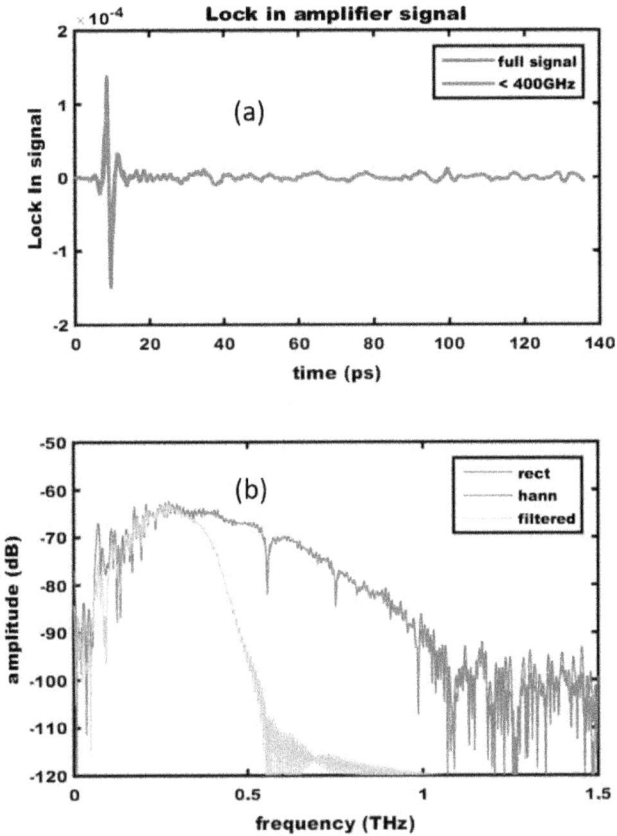

Figure 5.44: (a)The Lock in signal showing terahertz transient obtained without the filter on the THz path. (b) The Fast Fourier transform of the THz pulse giving the frequency spectrum of the THz signal without the filter(measurement obtained by OSCAT system).

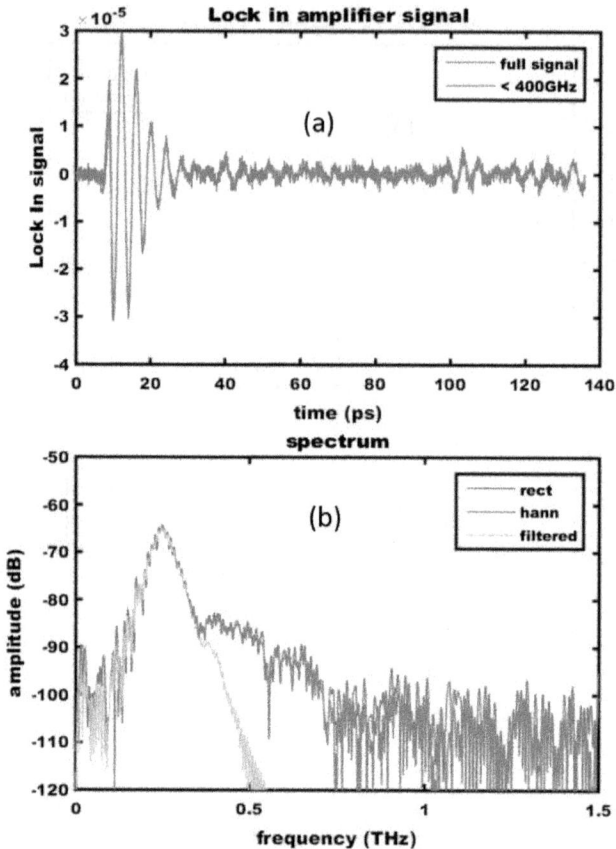

Figure 5.45: (a)The Lock in signal showing terahertz transient obtained with the filter on the THz path. (b) The Fast Fourier transform giving the frequency spectrum of the THz signal with the filter (measurement obtained by OSCAT system).

THz metrology

A second potential application of our system is in THz metrology. To demonstrate the potential applicability of the developed system in THz metrology, a 3.1 mm polyethylene block sample was analyzed for its thickness measurements. We demonstrate a non-destructive, contact free approach to determine the thickness of sample using the generated multifrequency CW THz source by precisely fine tuning the beat frequencies passing through the sample by adjusting the heater current of the diode in the range 0 to 350 mA. A multi-frequency THz map of the different THz frequency signals was obtained to determine the thickness of sample directly

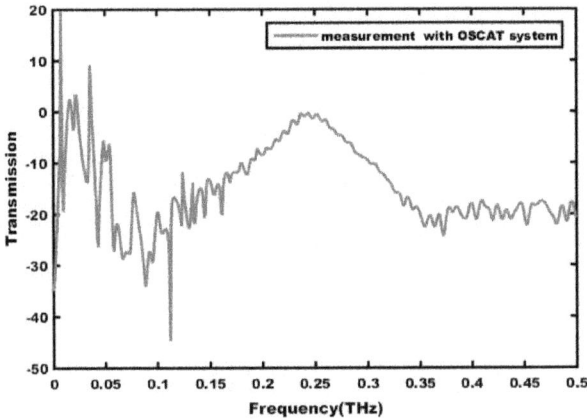

Figure 5.46: Transmission characteristics of the filter analyzed with the OSCAT system with peak transmittance centred around 250 GHz.

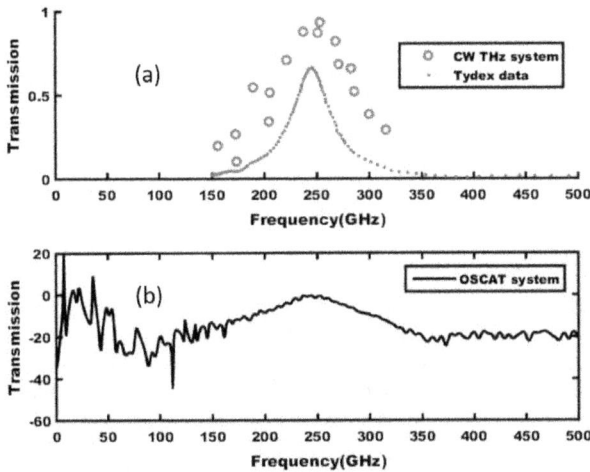

Figure 5.47: Transmission characteristics of the THz band pass filter measurement obtained with our CW THz system and compared with the Tydex data Figure(a) and OSCAT system measurement Figure(b).

from the position shift of the modes due to the delay introduced by the sample with respect to original mode position. The measurement was performed in the frequency range between 100 GHz to 315 GHz when the heater current was varied in steps of 20 to the maximum resistor heater voltage of 1800 mV which is equivalent to 350 mA heater injection current. A THz transmission based geometry was used while taking this thickness measurement, with the sample placed in between the transmitter and the receiving antenna. Figure 5.48 (a) shows a reference measurement taken without the sample and different THz frequencies were mapped showing their mode positions with increasing heater currents. Figure 5.48 (b) shows a measurement obtained when the sample was inserted on the THz beam path. The introduced shift of the position of the different THz frequency modes in comparison to the original mode position of the reference measurement are observed in the sample measurement. The shift in position of the modes is attributed to the refractive index and thickness of the sample which induces a delay to the THz signal before it arrives at the receiving antenna for detection.

From the multi-frequency THz color map Figure 5.48, a correlation coefficient was evaluated to determine the maximum intensity distribution at different position with and without the sample with respect to the specific heater current. This evaluation was based on the correlation coefficient relation $X(i, j)$ between the two images in Figure 5.48 given by

$$X(i, j) = corrcoef(X_1(:, i), X_2(:, j)) \qquad (5.24)$$

where $X_1(:, i)$ represents the image of the THz color map without the sample (reference measurement) while $X_2(:, j)$ represents the image of the THz color map obtained with the sample inserted in the THz path, i and j represents the different modes of THz signal in the THz color maps. Figure 5.49 illustrates the correlation coefficient of the reference measurement and the sample measurement showing the intensity distribution at different position in relation to the heater currents of the diode. The large correlation coefficient values occurs along the diagonal line pattern providing the maximum intensity distribution of the reference measurement, which shifts in position in the sample measurement due the refractive index as well as the thickness of the sample. To determine the sample thickness from the calculated correlation coefficients, the positions of the highest intensity with and without the sample for a particular heater current was evaluated as shown in Figure 5.50. The difference in the position of the maximum intensity shift for a particular heater current and corresponding position of the sample measurement provides the optical thickness of the sample. For example, the measured shift at 157 mA heater current, gives the maximum intensity position of the reference measurement at 157, while the corresponding intensity position for the sample measurements was 63, these values gives as a position difference of Δl=94 mm. The multi-frequency THz color map measurements were obtained by varying the delay stage with step size of 0.03, by multiplying this factor with the obtained difference value gives a corresponding optical thickness of 2.82 mm. The thickness obtained for different heater currents are summarized in table 5.1. The average optical thickness of the analyzed sample is 2.86 mm which is very close to the actual measurement of the sample which was obtained by the Vernier caliper of 3.1 mm.

The measurements obtained here were obtained by scanning the delay stage. At fixed delay stage we believe by only scanning the heater current the delay introduced by the sample would be sufficient to enable the thickness measurement of the sample to be determined. This would make the measurement with this system even more simplified, faster and more compact, thereby enabling measurement to be done without necessarily scanning the delay stage when making the reference measurement. We attempted to investigate this by scanning the heater current alone and

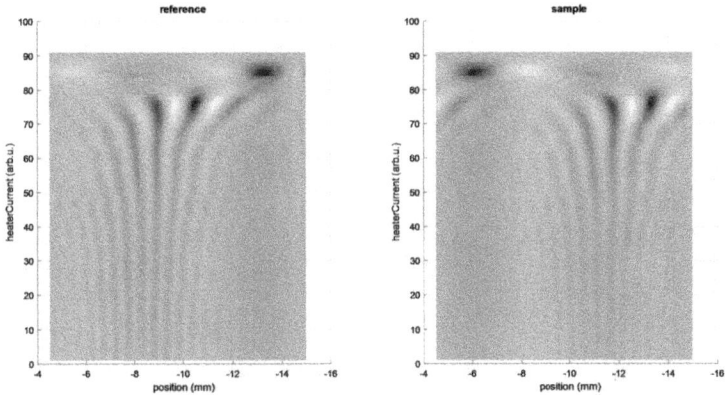

Figure 5.48: Multi-frequency THz signal color map (a) without the sample (reference) and (b) with the sample showing the shift of the modes position as a result of the insertion of the sample in the THz signal path.

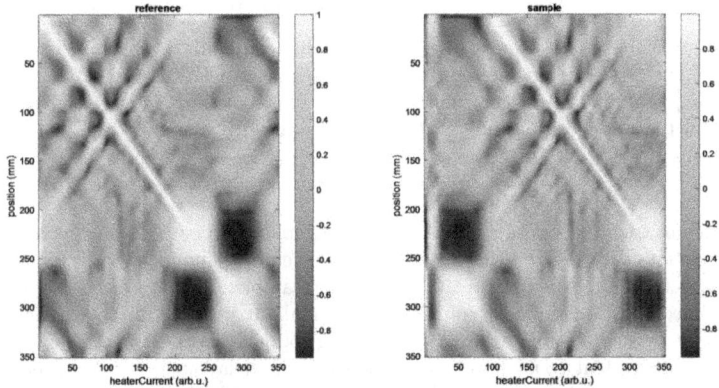

Figure 5.49: The correlation coefficient between the sample and the reference showing the maximum intensity shift upon insertion of the sample on the THz beam path.

from the obtained optical beat signals a reference multi-frequency THz map was obtained through numerical simulation. As a proof of the idea, Carsten Brenner did a simulation to determine the reference multi-frequency THz color map for the different mode positions from the optical spectrum measurement upon scanning the heater current from 0 mA to 350 mA. The simulated reference measurement was then employed to determine the thickness of the sample, however we suppose due to the power differences between simulated reference signal and the experimental measurements the determination of the correlation coefficient of the reference (simulated

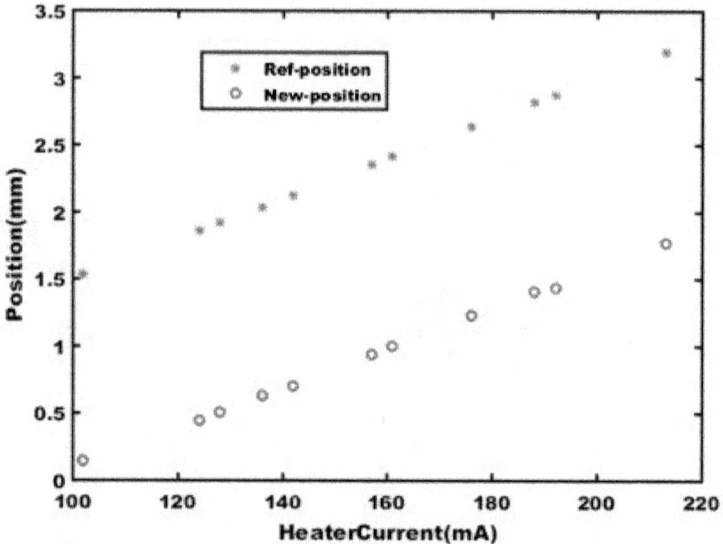

Figure 5.50: Positions of the maximum intensity plotted against heater current for the reference position and corresponding new position measurements with the sample at the same heater currents. The difference between the reference position and the new position due to the shift introduced by sample gives the thickness of the sample.

signal) and the sample THz signal measurement did not yield expected outcome. The intensity shift on the correlation coefficient measurement was expected to help in determining the optical thickness of the sample. This approach could have simplified the measurement complexity and make it possible for the measurement to be perfomed at fixed delay stage resulting to much shorter time in taking the measurement. If it could have been successful, the delay stage which introduces more measurement time, large size and complexity could have been eliminated. Figure 5.51 (a) shows a contour plot of the different modes obtained by scanning the heater current from 0 to 350 mA with the modes observed to overlab at higher heater current. Figure 5.51 (b) reference(experimental) multi-frequency THz color map measurement obtained by scanning the delay stage and changing the heater current in Figure 5.51 (c) the simulated optical modes are obtained by scanning the heater current without the delay stage being used and finally Figure 5.51 (d) illustrates the simulated reference multi-frequency THz color map obtained by just scanning the heater current without the involvement of the delay stage. From this measurement we observe that with only adjusting the heater current a reference multi-frequency THz color map can be obtained from tuning the optical beat signal similar to when the delay stage is employed. In this way the measurement could be simplified and could even be faster since its only sample measurement which could be taken while the reference measurement could be obtained from the optical spectrum measure-

Heater current	reference position	new position	difference	thickness(mm)
157	157	63	94	2.82
213	215	118	97	2.91
245	245	150	95	2.85
256	257	161	96	2.88
306	305	211	94	2.82

Table 5.1: Table of thickness measurement of the sample at different heater currents

ment by scanning the heater current. The simulated reference multi-frequency THz signal was obtained from the optical spectra based on the difference frequency equation given by

$$\Delta v_{ij} = \frac{c\Delta\lambda_{ij}}{\lambda_i\lambda_j} \tag{5.25}$$

where λ_i and λ_j represents the left mode and right mode at different heater currents, $\Delta\lambda_{ij} = \lambda_i - \lambda_j$ and c is the speed of light. The simulated THz fields $E_{sim}(t) \propto I_i.I_j$ where I_i and I_j are the respective intensities of the different frequency modes and the overall intensity of the simulated reference multifrequency THz signals will depend on all the difference frequency components power which can be described by

$$P_{ij} = P_i + P_j + 2\sqrt{mP_iP_j}cos(2\pi.v_{ij}t + \phi) \tag{5.26}$$

where P_i and P_j represents the respective powers for the left mode and right mode at different heater currents. With subscripts i and j being a positive integers.

To apply the simulated reference multi-frequency THz signals in thickness determination, the intensity correlation coefficient between reference signals, simulated signals and sample signals was examined. Figures 5.52 (a) gives the intensity correlation coefficient of the reference signals obtained when the delay stage was scanned. In Figures 5.52(b) the intensity correlation coefficient was obtained when the sample was introduced into the THz signal path with the delay stage scanned and different beat signals obtained by changing the heater current, while in Figure 5.52 (c) the intensity distribution of the reference signals is correlated to the signals of the sample measurement. In the Figures 5.52 (a)and (b), the measurement indicates the maximum intensity amplitude distribution occurs along the diagonal. For Figure 5.52 (c) the sample induces a shift of the maximum intensity in comparison to its original position. A comparison of the different heater current position and corresponding heater current position in the reference measurements, provides the difference between the two position of the heater current which gives the optical thickness of the sample. In Figure 5.53(a) correlation coefficient of the simulated reference signals made without scanning delay stage is shown. Figure 5.53(b)shows the correlation coefficient of reference signals obtained by scanning the delay stage and changing the heater current. Figure 5.53(c) shows the correlation coefficient between the simulated signals and reference measurement. In this case Figure 5.53(a) represents the maximum intensity along the diagonal for the simulated signals. Similarly, in Figure 5.53 (b) the maxima intensity of the reference occurs along the diagonal line. The corresponding correlation coffiecient between the simulated signals and reference signals gives plured intensity with the cross like pattern occurring around the position [200,250] in Figure 5.53 (c). Finally, in Figure 5.54 (a) a correlation

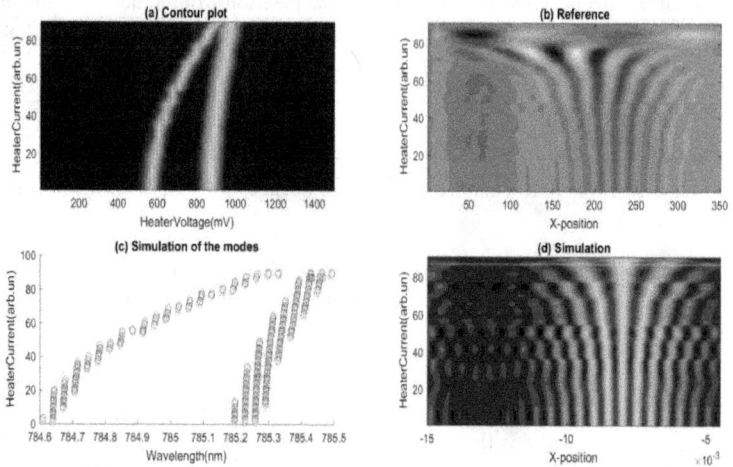

Figure 5.51: (a)contour plot of the different modes obtained by scanning the heater current from 0 to 350 mA(b)reference multifrequency THz color map measurement obtained by scanning the delay stage and changing the heater current(c)The simulated optical modes obtained by scanning the heater current(d)The simulated reference multifrequency THz color map obtained by scanning the heater current

coefficient of the simulated reference signal made without scaning the delay stage is shown while in Figure 5.54 (b) the correlation coefficient of sample measurement and the correlation coefficient between the simulated signals and sample measurement is obtained in Figure 5.54 (c). In this case also the maximum intensities occurs along the diagonal in Figure 5.54(a) and (b), for Figure 5.54(c) the cross like pattern where the maximum intensity occurs shifts to a new position [220,105], due to the delay introduced by the sample. We believe by optimizing the simulation parameters based on the optical spectrum power, correlation coefficient of signal with higher powers could give a clearer maximum intensity distribution which will enable the determination of the sample thickness directly from the shift introduced by the sample on the intensity. The results demonstrated here clearly shows that the developed system has potential in THz metrology in industrial thickness measurements.

5.9 Conclusion

In this chapter, we have discussed in section 5.8.1 the implementation of a compact terahertz system based on a two color DBR laser diode. The two color DBR laser diode is implemented using two DBR grating at the rear end which are monolithically integrated with a Y-shaped waveguide to realize a common output. This device was employed to excite photoconductive antennas to generate 300 GHz THz signal. The generated THz power was estimated to be in the low μW range, which was estimated from the photocurrent measurents. The difference frequency of the

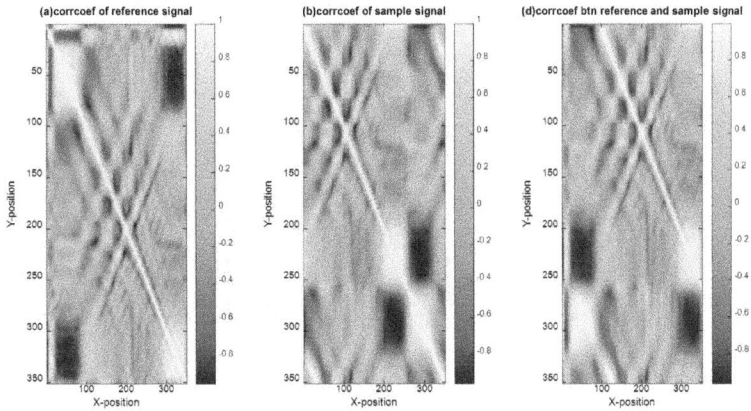

Figure 5.52: (a)correlation coefficient of the reference signal made with scaning delay stage and changing the heater current(b)the correlation coefficient of sample THz color map measurement obtained by scanning the delay stage and changing the heater current(c)The correlation coefficient between the reference signal and sample measurement

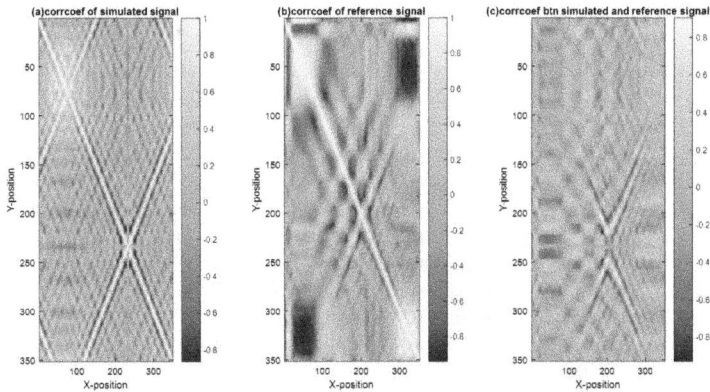

Figure 5.53: (a)correlation coefficient of the simulated reference signal made without scaning delay stage(b)the correlation coefficient of reference THz color map measurement obtained by scanning the delay stage and changing the heater current(c)The correlation coefficient between the simulated signal and reference measurement

demonstrated source could be adjusted from 280 GHz to 320 GHz by changing the injection current and temperature of the entire laser chip and the laser sub-mount simultaneously. This roughly can provide an approximately 40 GHz bandwidth. In brief, we demonstrated here a compact continuous wave THz system based on mono-

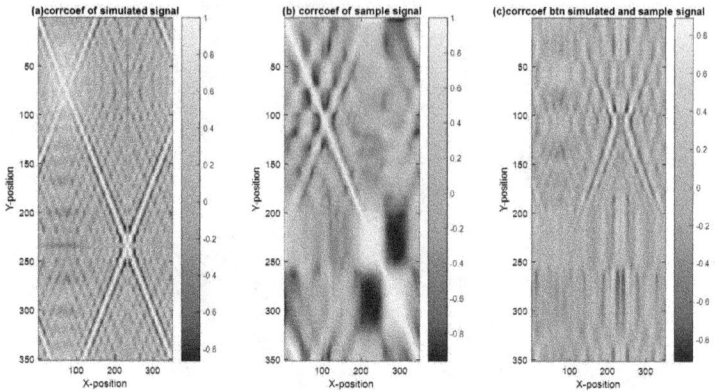

Figure 5.54: (a)correlation coefficient of the simulated reference signal made without scaning delay stage(b)the correlation coefficient of sample THz color map measurement obtained by scanning the delay stage and changing the heater current(c)The correlation coefficient between the simulated signal and reference measurement

lithic two color laser diode and explored its potential applications in non-invasive moisture measurements on drying pieces of paper and on leaf under drought induced stress. The changes of THz transmission through the pieces of papers and leaf correlate with the moisture content in the samples. THz radiation is highly absorptive in moisture, which could actually be used to sense even low water content. This makes it a valuable tool for plant scientists for inspection of water content in plants tissues for timely irrigation before wilting of the plants. The results obtained demonstrate a stable, compact and cost-effective CW THz source with potential use in industrial non-destructive testing measurements.

In section 5.8.6, we have discussed a new DBR two-color laser device which is electrically tunable. The spectral separation of the two modes is adjusted via carrier injection by changing the current injection to the resistor heaters implemented in the DBR section of the diode between 0 to 320 mA. Each laser could be independently tuned. The tuning range of laser 1 was from 784.4 nm to 785.6 nm with current injection to micro heater adjusted between 0 to 350 mA this gives a wavelength span of 1.2 nm. While laser 2, under similar injection current was tuned from 785.0 nm to 785.4 nm which gives a wavelength difference of 0.4 nm. The optical beat tuning range of the demonstrated structure is between 100 GHz - 300 GHz; only limited by the mode overlap at higher resistor heater currents between 250 mA to 350 mA. The structure generated multiple frequencies due to the observed lateral modes besides the two main modes. The generated THz radiation was detected in a homodyne setup. Finally, a simple spectrometer system suitable for THz metrology measurements such as thickness determination of Polyethyrene sample (PE) was realized and also its application in spectroscopy was demonstrated by the determination of the spectroscopic transmission characteristics of a THz filter.

Chapter 6

Summary and conclusion

In summary, a brief introduction of this dissertation is given in chapter 1. The fundamentals of semiconductor diode lasers are highlighted in the subsequent chapter 2, with brief discussion on the types of diode lasers used in this study given. In chapter 3, we introduce the various methods used for THz detection and point out the need to develop compact THz detection schemes in order to bring THz technology into many applications. In chapter 4, the investigation of the interaction of THz signal from various sources with a semiconductor diode laser is discussed, with particular focus to demonstrate detection of THz radiation with diode lasers. Our expectation was that the detection would have taken place in the carrier plasma. Since the carrier plasma in a diode laser responds to THz frequencies, we hoped that it would also responds directly to THz radiation injected into the diode laser. As a result, THz modulation of the optical gain in the laser will occur and generate sidebands in the NIR optical spectra. Thus, the injected THz signal will be converted into the NIR spectral regime and could be easily analyzed with an optical spectrometer. Unfortunately, we didn't observe the expected results due to various reasons such as:

- The SynView THz source: THz radiation emited with the SynView THz source was about 60 μW, such low power could be insufficient and might have limited the interaction between THz radiation and diode signal. Also, due to the high likelihood of losses during coupling into the p-n junction, very low signal would have been available for interaction process. Besides that, the spot size of the SynView source THz signal was large and therefore quite challenging to focus it into the diode laser.

- Gunn diode: Despite the fact it emited more optical power of 30 mW, its spectral spot size was broad, since it emits at lower THz frequency of 120 GHz. Coupling such a signal with broad spot size into the diode active region is quite challenging. Even after incorporating some waveguide, the signal power coupled into the diode it appeared was not enough; hence no observable interaction results were obtained.

- THz TDs system: The flunctuations of the diode signal during the interaction experiment become dominant and made it more challenging to take measurements. Also, since this was a pulsed system and the operation of the diode was in continuous mode the instant of interaction between THz signal and diode signal could have occured at a very fast time instance which could have been difficulty to be noticed.

In chapter 5, we have discussed the development of a compact terahertz source with a two color monolithically integrated DBR laser diode. The two color Y-shaped DBR laser source discussed here clearly shows that is suitable for THz frequency generation in the range of 300 GHz. The generated THz power was estimated (from the photocurrent measurements) to be in μW range. The tuning of the difference frequency from 286 GHz to 320 GHz was done by changing the injection current and temperature of the entire laser and the submount simultaneously. An approximately 40 GHz bandwidth can be demonstrated with this device. Moreover, the potential applications for the developed system in non-invasive moisture measurements on drying pieces of paper and on moisture monitering in a plant leaf under drought induced stress was explored. The changes of THz transmission through the pieces of papers and leaf correlate with the moisture content in the plant leaf and the pieces of paper samples undergoing drying process. THz radiation is highly absorptive in moisture which could actually be used to sense even low water content. This makes it a valuable tool for plant scientists for inspection of water content in plants tissues for timely irrigation before wilting. Furthermore, the obtained results have clearly demonstrated that the developed CW THz source can be exploited in industrial non-destructive testing measurements.

Finally, in chapter 5, we discussed the development of another compact CW terahertz source, based on an electrically tunable DBR monolithically intergrated diode laser. The latter is electrically tunable and is designed with capability to adjust the optical beat frequency of the emitted wavelengths. Optical beat tuning is accomplished with micro resistor heater implemented on top of the DBR section. It changes the local temperature of the DBR segment via carrier injection thereby shifting the emission wavelength as a result of the changes in the refractive index. We demonstrated terahertz wave generation in the frequency range of 100 GHz-300 GHz. This frequency band was only limited due to mode overlap at higher resistor currents which resulted to zero difference frequency between the two modes. Furthermore, we demonstrated the potential use of the implemented spectrometer in non-contact measurement for the thickness determination of a sample of PE block and the spectroscopic transmission characteristics of a THz filter.

In conclusion, we didn't manage to achieve a semiconductor diode laser based THz detection by interacting THz radiation with diode laser but, we believe if more investigation could be carried out there is likelihood of realizing such a compact and cost effective detection scheme. We recommend that powerful CW THz source such as THz gas laser(or any other CW source) which can deliver intense THz signal be used. In such investigation the laser should be operated in CW mode. Silicon lenses should be incorporated to enable small spot size of THz frequencies is achieved therefore enhancing the coupling of THz signal into the active region of the diode. In the case pulsed THz source is used, the diode should be stabilized and if possible shielded from any random interference of the signal from the surrounding instruments as well its injection current to be modulated to match the repetition rate of the pulsed THz source. Finally, in the case of the eDBR laser we recommend more investigation to be done on possibility of using the device for near infrared spectroscopy since each laser wavelength could be tuned independently. For the generated multifrequency CW THz signal we recommend, more research to be done in order to realize its full application pontential more so its suitability in THz imaging and gas spectroscopy.

Bibliography

[1] M. C. Hoffmann and J.A Fulop, Intense ultrashort terahertz pulses: Generation and application, *Journal of physics D: Appl. phys.* 44, 83001,2011.

[2] I. Park, "Investigations of the Generation of Tunable Continuous-Wave Terahertz Radiation and Its Spectroscopic Applications," *PhD Thesis Darmstadt University of Technology*, 2007.

[3] C. Brenner, S. Hoffmann, M.R. Hofmann, M. Salhi, M. Koch, A. Klehr, G. Erbert, and G. Tränkle,Detection of THz radiation with diode lasers, *Electron. Lett.* ,43,16,2007.

[4] C. Brenner, S. Hoffmann, M.R. Hofmann, M. Salhi, M. Koch, A. Klehr, G. Erbert, G. Tränkle, J.T. Steiner, M. Kira and S.W. Koch, Detection of THz radiation with semiconductor diode lasers, *Appl. Phys. Lett.* 91, 101107 ,2007.

[5] C. Brenner, S. Hoffmann, M.R. Hofmann, Interaction of Semiconductor Laser Dynamics with THz Radiation, *Advances in Solid State Physics,* 47,179,2008.

[6] S.Ragam, CW THz wave generation system with diode laser pumping, *www.intechopen.com*, April 25,2012.

[7] A.Lisauskas,M.Dias,S.Belz,H.G.Roskos and M.Feiginov, Concept of internal mixing in semiconductor lasers and optical amplifiers for room temperature generation of tunable continues terahertz waves, *Journal of nanoelectronics and optoelec* 2,1,2007.

[8] C. Brenner, S. Hoffmann, C.-S. Friedrich, T. Schlauch, A. Klehr, G. Erbert, G. Tränkle, C. Jördens, M. Salhi, M. Koch and M. R. Hofmann, Semiconductor laser based THz generation and detection, *Phys. stat. sol.* ,C6,564,2009.

[9] S. Hoffmann, X. Luo and M.R. Hofmann, Bandwidth limitations of two-colour diode lasers for direct terahertz emission,*Electronic lett.*.42,696,2006.

[10] S. Hoffmann, M. Hofmann, E. Bründermann, M. Havenith, M. Matus, J.V. Moloney, A.S. Moskalenko, M. Kira, S.W. Koch, S. Saito and K. Sakai, Four-wave mixing and direct terahertz emission with two-color semiconductor lasers, *Appl. Phys. Lett.* 84, 3585, 2004.

[11] R.N.Hall, G.E.Fenner,J.D.kingslev, T.J.soltys and R.O Carlson,'Coherent light emission from GaAs junctions', *Phys. Rev.Lett.*9,366,1962.

[12] S.Hoffmann and M.R. Hofmann, Generation of Terahertz radiation with two color semiconductor lasers,*Laser and Photon.Rev*,1,44 ,2007.

[13] M.Maiwald, B. Sumpf and G. Tränkle,'Rapid and adjustable shifted excitation Raman difference spectroscopy using a dual-wavelength diode laser at 785 nm', *J Raman Spectrosc.*,1-11,2018.

[14] M.N. Cherkashin, C.Brenner, L.Göring, B. Döpke, N. C. Gerhardt, M. R. Hofmann , Laser diode based photoacoustic setup to analyze Grüneisen relaxation effect induced signal enhancement, *Proc. SPIE 9539, OptoAcoustic Methods and Applications in Biophotonics II*, 95390M ,Munich, 2015.

[15] A. Müller, S. Marschall, O.B. Jensen, J. Fricke, H. Wenzel, B.Sumpf, and P. E. Andersen,Diode laser based light sources for biomedical applications, *Laser Photonics Rev.* 7, 5, 605–627,2013.

[16] O.B.Jensen,A.K.Hansen,A.Müller,B.Sumpf,A.Unterhuber,W.Drexler,P.M.Petersen, and P.E.Andersen,Power Scaling of Nonlinear Frequency Converted Tapered Diode Lasers for Biophotonics, *IEEE J. Sel. Top. Quantum Electron.* 20, 2, 7100515,2014.

[17] *www.phys.unm.edu/msbahae/physics464/semiconductor lasers.pdf* accessed on 02.Jan,2019.

[18] L.G. Johansen, "Radiation hardsilicon microstrip detectors for use in ATLAS at CERN," *PhD dissertatin University of Bergen*,Norway 2005.

[19] M.Pospiech and S. Liu, "Laser diodes an introduction," *University of Hannover*,2004.

[20] P.W.Epperlein,Basic Diode laser Engineering principles,Semiconductor Laser Engineering, Reliability and Diagnostics: A Practical Approach to High Power and Single Mode Devices,*John wiley and sons*,2013.

[21] W.Demdröder, "Laser spectroscopy:Basic principals," *Springer*,1,4,pp. 314-318, 2008.

[22] T.A. Heumier, "Mode hopping in semiconductor lasers," *PhD Dissertation, Montana State University*,1992.

[23] Z.I.Alferov,Double heterostructure lasers:early days and future perspectives,*IEEE J.of sel.Topics in Quant.Electro*.6,832-840,2000.

[24] P.Blood, Quantum confined laser devices(optical gain and recombination in semiconductors), New York: *Oxford University Press*,2015.

[25] S.O. Kasap, Optoelectronics,*Prentice Hall*,1999.

[26] H.Tulsani,electronic devices, *edetec106.blogspot.com*,2016

[27] T.Numai,Fundamentals of Semiconductor Lasers, *Springer Series in Optical Sciences* 93,23-25, 2015

[28] B. Sumpf, M. Maiwald, A. Müller, J. Fricke, P. Ressel, F. Bugge, G. Erbert and G. Tränkle,, "Comparison of two concepts for dual wavelength DBR ridge waveguide diode lasers at 785 nm suitable for shifted excitation Raman difference spectroscopy'," *Appl. Phys. B* , 120,2, 261-269, 2015.

[29] M. Maiwald, B. Eppich, J. Fricke, A. Ginolas, F. Bugge, A. Klehr, B. Sumpf, G. Erbert, G. Tränkle, "Shifted Excitation Raman Difference Spectroscopy using a Dual-Wavelength DBR Diode Laser at 785 nm," *Proc. of SPIE*,9313, 93130Y, 2015.

[30] D.Jeannotte, External cavity with Littrow-Mounted diffraction grating, Rochester, New York: *University of Rochster*,1994.

[31] C.S. Friedrich, C. Brenner, S. Hoffmann, M. R. Hofmann ,A. Schmitz,I. Camara Mayorga, A. Klehr and G. Erbert,, "New Two-Color Laser Concepts for THz Generation," *IEEE J. of Sel. Topics in Quantum Electron*,14,2, 2008.

[32] A. Rogalski, F. Sizov,Terahertz detectors and focal plane arrays, *Optoelectronics rev.*,19(3),346–404,2011.

[33] M. Tonouchi,Cutting-edge terahertz technology, *Natur. Photon.* ,2007.

[34] R. Wilk,Continuous wave Terahertz spectrometer with coherent detection, *Bull. Pol. Ac.: Tech.*54(4),2010.

[35] R. Guo,T. Ikar,J. Zhang, H. Minamide,and H.Ito,Frequency-agile THz-wave generation and detection system using nonlinear frequency conversion at room temperature, *Opt. Express*18(16)16430-16436,2010.

[36] Y. S. Lee, Principles of Terahertz Science and Technology, *Springer Science*, New York ,2009

[37] E.Bründermann ,H-W. Hübers and M.F Kimmitt, "Terahertz techniques," *Springer series in optical sciences*, 2012.

[38] A. Rogalski , "Infrared detectors: status and trends, *Progress in Quantum Electronics* , 27, 59-210, 2003.

[39] F.Hindle,C.Yang,G. Mouret, A.Cuisset, R. Bocquet ,J.F Lampin, K. Blary, E. Peytavit, T. Akalin and G. Ducournau, "Recent Developments of an Opto-Electronic THz Spectrometer for High-Resolution Spectroscopy, *sensors*9, 9039-9057,2009.

[40] D. H. Auston, K. P. Cheung, and P. R. Smith, Picosecond photoconducting Hertzian dipoles, *Appl. Phys. Lett.*, 45, 284 ,1984.

[41] Y. Cai, I. B. Lopata, J. Wynn,L.Pfeier, J.B Stark, Q.Wu, ,X.C Zhang and J.F Federici, Coherent terahertz radiation detection: Direct comparison between free-space electro-optic sampling and antenna detection," *Appl.Phys.Lett* 73, 444-446 ,1998.

[42] A. Leitenstorfer, S. Hunsche, J. Shah, M. C. Nuss, and W. H. Knox, Detectors and sources for ultrabroadband electro-optic sampling: experiment and theory, *Appl. Phys. Lett.*, 74, 1516 ,1999

[43] G. Gallot, and D. Grischkowsky,Electro-optic detection of terahertz radiation," *J. Opt. Soc. Am. B* 16,1204-1212 1999.

[44] W. Qiao, D. Stephan, M. Hasselbeck, Q. Liang, and T. Dekorsy, Low- temperature THz time domain waveguide spectrometer with butt-coupled emitter and detector crystal, *Opt. Express*.20 (18), 19769–19777,2012

[45] C. Winnewisser, P. U. Jepsen, M. Schall, V. Schyja, and H. Helm, "Electrooptic detection of THz radiation in LiTaO3, LiNbO3 and ZnTe, *Appl. Phys. Lett.*, 70, 3069 ,1997

[46] R.Ulbricht, E. Hendry,J. Shan, T.F Heinz, and M.Bonn, Carrier dynamics in semiconductors studied with time-resolved terahertz spectroscopy, *Rev. Mod. Phys*,83, 2, 2011

[47] J-F.Lampin, L.Desplanque, and F.Mollot,Electro-absorption sampling at terahertz frequencies in III-V semiconductors," *C.R.Physique* 9, 153-160 ,2008.

[48] S. Hoffmann, M. Hofmann, E. Bründermann, M. Havenith, M. Matus, J.V. Moloney, A.S. Moskalenko, M. Kira, S.W. Koch, S. Saito and K. Sakai, Four-wave mixing and direct terahertz emission with two-color semiconductor lasers, *Appl. Phys. Lett.* 84, 3585,2004.

[49] M. Germano,A, Massimo, C. Virginie, J. Mojca, G. Peter,Running electric field gratings for detection of coherent radiation, *Journ. of the Optical Society of America B.*, 32. 1078. 10.1364/JOSAB.32.001078.2015.

[50] S. Houver, A. Lebreton, R. Colombelli, L. Li, E. H. Linfield, A. G. Davies,J. Tignon,S. S. Dhillon,Nonlinear Frequency Mixing in Quantum Cascade Lasers:Towards Broadband Wavelength Shifting and THz Up-Conversion, *IEEE publish.*,2016.

[51] Y. Takida, K. Nawata,S. Suzuki, M. Asada,H. Minamide, Nonlinear optical detection of terahertz-wave radiation from resonant tunneling diodes, *Opt. Express*. ,25,5389-5395,2017.

[52] J. Madéo, P. Cavalié,J.R. Freeman,N. Jukam,J. Maysonnave,K. Maussang,H. E. Beere, D. A. Ritchie,C. Sirtori,J. Tignon, S. S. Dhillon,All-optical wavelength shifting in a semiconductor laser using resonant nonlinearities,*Nature Photonics*,2012.

[53] M.G. Littman, Single mode operation of grazing incidence pulsed dye laser, *Opt. Lett*.3 138 ,1978.

[54] C. am Weg, W. von Spiegel, R. Henneberger, R. Zimmermann, T. Loeffler,H. G. Roskos, "Fast active THz camera with range detection by frequency modulation," *Proc. SPIE 7215, Terahertz Technology and Applications II*,72150F,2009.

[55] H. Hirori,A.Doi, F.Blanchard, and K.Tanaka, Single-cycle terahertz pulses with amplitudes exceeding 1MV/cm generated by optical rectification in $LiNbO_3$, *Appl. Phys. Lett.* 98,091106,2011.

[56] M.C Hoffmann and J.A Fülöp , "Intense ultrashort terahertz pulses: generation and applications," *J. Phys. D: Appl. Phys.*, 44083001, 2011.

[57] P.U. Jepsen, D. G. Cooke, and M. Koch , "Terahertz spectroscopy and imaging–Modern techniques and applications.," *Laser Photonics Rev,* , 5, 1, 124-166, 2011.

[58] Y.D.Hsieh, S.Nakamura, D.G.Abdelsalam, T.Minamikawa,Y. Mizutani, H.Yamamoto, T.Iwata, F. Hindle and T.Yasui, "Dynamic terahertz spectroscopy of gas molecules mixed with unwanted aerosol under atmospheric pressure using fibre-based asynchronous optical- sampling terahertz time domain spectroscopy," *Scientific Reports,* 6, 28114, 2016.

[59] B. Sartorius, M. Schlak, D. Stanze, H. Roehle, H. Künzel, D. Schmidt, H.-G. Bach, R.Kunkel, and M. Schell, , " Continuous wave terahertz systems exploiting 1.5μm telecom technologies,," *Opt. Express,*17, 17, 15002 , 2009.

[60] S. Jung, A. Jiang, Y. Jiang, K. Vijayraghavan, X. Wang, M. Troccoli and M. A. Belkin,, " Broadly tunable monolithic room-temperature terahertz quantum cascade laser sources," *Nature communic.* , 5267, 2014.

[61] I.S. Gregory,H.Page, and l.Spencer,'Continuous-wave terahertz photomixer systems for real-world applications ,*Springer,* 67–184.,2007

[62] M. Tani, P. Gu, M. Hyodo, K. Sakai, and T. Hidaka , "Generation of coherent terahertz radiation by photomixing of dual-mode lasers," *Opt. Quantum Electron.,* 32, 503-520, 2000.

[63] M.Uemukai and T.Suhara, "Integrated two-wavelength dbr lasers for tunable photomixing thz-wave generation," *Ecio-conference* , 2012.

[64] S Hoffmann,M Hofmann,M.Kira and S.W.Koch, "Two-colour diode lasers for generation of THz radiation," *Semicond. Sci. Technol.,* 20, S205-S210, 2005.

[65] Verghese, S., McIntosh, K. A., Calawa, S., Dinatale, W., Duerr, E. K., and Molvar, K. A., "Generation and detection of coherent terahertz waves using two photomixers," Appl. Phys. Lett 73, 38243826 (1998).," *Appl. Phys. Lett* ,73, 3824-3826, 1998.

[66] N. M. Burford, and M. O. El-Shenawee, "Review of terahertz photoconductive antenna technology,," *Optical Engineering* , 56, 1,010901, 2017.

[67] N. Kim, H-C Ryu, D. Lee, S-P Han, H. Ko, K. Moon, J-W Park,M. Y Jeon and K. H Park , "Monolithically integrated optical beat sources toward a single-chip broadband terahertz emitter," *Laser Phys.Lett,* 10, 085805, 2013.

[68] R.E Miles,X-C.Zhang,H.Eisele,A.Krotkus, "Terahertz frequency detection and identification of materials and objects," *Springer Science,* 173, 2007.

[69] P.Gu, M. Tani, M. Hyodo, K. Sakai and T. Hidaka, "Generation of cw-Terahertz Radiation Using a Two-Longitudinal-Mode Laser Diode," *Jpn. J. Appl. Phys.,*37, 976–978, 1998.

[70] N.Kim, S. Han, H.Ko,Y.A. Leem,H.Ryu,C.W. Lee, D. Lee, M.Y. Jeon, S. K. Noh, and K.H. Park, , "N.Kim, S. Han, H.Ko,Y.A. Leem,H.Ryu,C.W. Lee, D.

Lee, M.Tunable continuous-wave terahertz generation/detection with compact 1.55 μm detuned dual-mode laser diode and InGaAs based photomixer," *Opt. Express*,19, 16, 15399, 2011.

[71] N. Kim, J. Shin, E. Sim, C. W.Lee, D-S Yee, M. Y. Jeon, Y. Jang, and K. H. Park, "Monolithic dual-mode distributed feedback semiconductor laser for tunable continuous-wave terahertz generation,," *Opt. Express*,17,13851-13856, 2009.

[72] K.Moon, N. Kim,J-H Shin,Y-J Yoon, S-P Han, and K.H. Park, "Continuous-wave terahertz system based on a dual-mode laser for real-time non-contact measurement of thickness and conductivity," *Opt. Express*,22, 3, 2259-2266, 2014.

[73] N.Kim, S.Han, K.Moon, I.Lee, E.S.Lee, and K.H.Park, "Optical Characteristics of 1.3-μm Dual-Mode Laser Diode with Integrated Semiconductor Optical Amplifier," *OSA publishing*, 2014.

[74] S.-H. Yang, R. Watts, X. Li, N. Wang, V. Cojocaru, J. O'Gorman, L. P. Barry, and M. Jarrahi , "Tunable terahertz wave generation through a bimodal laser diode and plasmonic photomixer," *Opt. Express*,23, 24, 31206-31215, 2015.

[75] R. A. Mohandas, J.R. Freeman,M.Natrelle,M.C. Rosamond,L. Ponnampalam, M.J. Ficealwyn, J. Seeds, P. J. Cannard,M. J. Robertson,D. G. Moodie,A. G. Davies,EH. Linfield,and P. Dean, Terahertz generation mechanism in nanograting electrode photomixers on Fe-doped InGaAsP, *Opt. Express* 25,9,2017.

[76] H Schmitz, Optimization of a continuous-wave THz spectrometer for coherent broadband solid-state spectroscopy, *PhD thesis,Universität zu Köln*, 2012.

[77] E. Castro-Camus and M.Alfaro, "Photoconductive devices for terahertz pulsed spectroscopy:a review(invited), *Photon. Res.*, 2016

[78] M. Venkatesh, K.S. Rao, T.S. Abhilash, S.P. Tewari, A.K. Chaudhary, "Optical characterization of GaAs photoconductive antennas for efficient generation and detection of Terahertz radiation, *Photonic materials* 36, 596–601, 2014

[79] X-C.Zhang and J.Xu, "Introduction to THz wave Photonics," London, *Springer science*, 2010

[80] I. C. Mayorga, E. A. Michael, A. Schmitz, P. van der Wal, R. Güsten, K. Maier, and A. Dewald, "Terahertz photomixing in high energy oxygen- and nitrogen-ion-implanted GaAs," *Appl. Phys. Lett.*, 91,031107, 2007

[81] H.C Ryu,N. Kim,S.Han, H.Ko,J-W. Park,K.Moon and K.H Park, "Simple and cost-effective thickness measurement terahertz system based on a compact 1.55μm /4 phase-shifted dual-mode laser," *Opt. Express*, 20, 23, 25990-25999, 2012

[82] F. Eichhorn, Fiber laser based broadband THz imaging systems, *Technical University of Denmark*, 2009.

[83] W. Withayachumnankul, Engineering aspects of Terahertz time-domain spectroscopy, *The University of Adelaide*, Australia, 2009.

[84] N. M. Burford and M. O.El-Shenawee, Review of terahertz photoconductive antenna technology, *Optical Engineering*, 56, 1, 010901, 2017.

[85] S. Lepeshov, A. Gorodetsky, A. Krasnok, E. Rafailov,and P. Belov, "Enhancement of terahertz photoconductive antenna operation by optical nanoantennas, *Laser Photonics Rev.* , 11, 1, 1600199, 2017.

[86] Z.Piao,M.Tani and K.Sakai, "Carrier dynamics and terahertz radiation in photoconductive antennas," *Jpn.J.App.Phys*, 39, 96-100, 2000.

[87] P. Uhd Jepsen, R. H. Jacobsen, and S. R. Keiding, "Generation and detection of terahertz pulses from biased semiconductor antennas, *J. Opt. Soc. Am. B*, 13, 11, 2424-2428, 1996.

[88] M. Tani, Y. Hirota, C. T. Que, S. Tanaka, R. Hattori, M. Yamaguchi, S. Nishizawa, and M. Hangyo. , "Novel terahertz photoconductive antennas., *Int. J. Infrared Millimeter waves*, 27, 531-546, 2006.

[89] M. Tani ,O. Morikawa,S. Matsuura,and M. Hangyo, "Generation of terahertz radiation by photomixing with dual and multiple-mode lasers, *Semicond. Sci. Technol.*, 20, S151–S163, 2005.

[90] T.Kleine-Ostmann, "Generation,Modulation and detektion von THz-Trägerwellen. *Diplomarbeit,Institut für Hochfrequenztechnik,TU Braunschweig*,2001.

[91] I Park, C Sydlo, I Fischer, W Elsäßer and H L Hartnagel, "Generation and spectroscopic application of tunable continuous-wave terahertz radiation using a dual-mode semiconductor laser," *Meas. Sci. Technol.*, 19, 065305, 2008.

[92] R.B. Kohlhaas,A.Rehn,S.Nellen,M.Koch, M. Schell,R. J. B. Dietz,AND J. C. Balzer,Terahertz quasi time-domain spectroscopy based on telecom technology for 1550 nm, *Opt. Express*25,11,12853,2017.

[93] G. Carpintero,L.Enrique, G. Munoz,H. L.Hartnagel,S. Preu and A. V.Räisänen, Semiconductor terahertz technology:Devices and systems at room temperature operation, *John Wiley and sons,Ltd,First edition*, 5-22,2015.

[94] M. Tani, P. Gu, M. Hyodo, K. Sakai, and T. Hidaka, "Generation of coherent terahertz radiation by photomixing of dual-mode lasers," *Optical and Quantum Electronics*, 32, 503-520, 2000.

[95] S. Matsuura, M.Tani and k. sakai, "Generation of coherent terahertz radiation by photomixing in dipole photoconductive antennas," *Appl.Phys.lett,* 70, 559, 1997.

[96] A. Krotkus, "Semiconductors for terahertz photonics applications," *J. Phys. D: Appl. Phys.* , 43, 273001, 2010.

[97] S. Hoffmann, M. Hofmann, M. Kira, and S.W. Koch,"two-color diode lasers for generation of THz radiation," *Semicon.Sci.Technol.*, 20,S205, 2005.

[98] E.R,Brown, K.A.McIntosh, F.W.Smith, M.J.Manfra and C.L. Dennis, "Measurements of optical-heterodyne conversion in low-temperature-grown GaAs," *Appl. Phys. Lett.* , 62,1206, 1993.

[99] K.A, McIntosh, E.R. Brown, K.B. Nichols, O.B. McMahon, W.F. DiNatale and T.M. Lyszczarz, "Terahertz photomixing with diode lasers in low-temperature-grown GaAs,," Appl.Phys. Lett. , 67,3844, 1995.

[100] M. Uemukai and T.Suhara, " Integrated two-wavelength DBR lasers for Tunable Photomixing THz-Wave generation," *European Conference of Integrated Optics*, 2012.

[101] P. Gu, F. Chang, M. Tani, K. Sakai and C-L Pan, "Generation of Coherent cw-Terahertz Radiation Using a Tunable Dual-Wavelength External Cavity Laser Diode," *Jpn. J. Appl. Phys.*,38,L1246-L1248, 1999.

[102] T. Kleine-Ostmann, P. Knobloch, M. Koch, S. Hoffmann, M. Breede, M. Hofmann, G. Hein, K. Pierz, M. Sperling, and K. Donhuijsen, , "Continuous-wave THz imaging," *Electron. Lett.*,37,24,1461, 2001.

[103] S. Hoffmann, M. Hofmann, "Generation of Terahertz with two-color semiconductor lasers,," *Laser and Photon. Rev*, 1,44-56, 2007.

[104] C. Brenner, S. Hoffmann, C.-S. Friedrich, T. Schlauch, A. Klehr, G.Erbert, G. Trnkle, C. Jrdens, M. Salhi, M. Koch and M. R. Hofmann , "Semiconductor laser based THz generation and detection," *Phys. stat. sol.*,C6,564, 2009.

[105] N. Kim, H-C Ryu, D Lee, S-P Han, H Ko, K Moon, J-W Park,M Y Jeon and K H Park, "Monolithically integrated optical beat sources toward a single-chip broadband terahertz emitter," *Laser Phys. Lett.* ,10,085805, 2013.

[106] S. Iio ,M. Suehiro ,T. Hirata and T. Hidaka, "Two-longitudinal-mode laser diodes," *IEEE Photonics*,9, 7, 1995.

[107] T. Hidaka, S. Matsuura, M. Tani, K. Sakai, "CW terahertz wave generation by photomixing using a two-longitudinal-mode laser diode," E*Electronic Lett.*,33,24,2039 - 2040, 1997.

[108] P. Gu, M. Tani, M. Hyodo, K. Sakai and T. Hidaka, "Generation of cw-Terahertz Radiation Using a Two-Longitudinal-Mode Laser Diode," *Jpn.J.Appl.Phys.*,37, L976-L978, 1998.

[109] M. Uemukai, H.Ishida, A. Ito, T.Suhara,H.Kitajima, A.Watanabe, and H. Kan, "Integrated AlGaAs Quantum-Well Ridge-Structure Two-Wavelength Distributed Bragg Reflector Laser for Terahertz Wave Generation," *Jpn. J. Appl. Phys.* , 51,020205, 2012 .

[110] K.Namje , H.Sang-Pil , K. Hyunsung, L. Y. Ahn , R. Han-Cheol, L. Chul , L. Donghun, J. Min yong, N. Sam and P. k. Hyun., "Tunable continuous-wave terahertz generation/detection with compact 1.55 nm detuned dual-mode laser diode and InGaAs based photomixer," *Opt. Express*, 19,16, 15397-15403, 2011.

[111] N.Kim and S.Han and K.Moon and I.Lee and E.S.Lee and K.H.Park, "Optical Characteristics of 1.3-μm Dual-Mode Laser Diode with Integrated Semiconductor Optical Amplifier," *OSA Publishing*, 2014.

[112] T.Gobel and D.Stanze and U.Troppenz and J.Kreissl and B.Sartorius and M.Schell, "Integrated continuous-wave THz control unit with 1 THz tuning range," *Int. Conf. Infrared, Millimeter, Terahertz Waves*, IRMMW-THz, 1-3, 2012.

[113] M.Sun and S.Tan and S.Liu and F.Guo and D.Lu and R.Zhang and Q.Kan and C.Ji, "Monolithically Integrated Two-Wavelength Distributed Bragg Reflector Laser for Terahertz Generation," *OSA Publishing*, 2016.

[114] M.Sun and S.Tan and S.Liu and F.Guo and D.Lu and R.Zhang and Q.Kan and C.Ji, "Continuously Tunable Terahertz Signal Generation with an Integrated 1.55-μm Dual-Wavelength DFB Photonic Chip," *IEEE Xplore,Photonics Conference (IPC)*, 2016.

[115] J. O. Gwaro, C. Brenner, B. Sumpf, A. Klehr, J. Fricke and M. R. Hofmann, "Terahertz frequency generation with monolithically integrated dual wavelength distributed Bragg reflector semiconductor laser diode," *IET Optoelectronics*, 11,2,49-52, 2017.

[116] J. O. Gwaro, C. Brenner, B. Sumpf, A. Klehr, J. Fricke and M. R. Hofmann, "Compact continuous wave THz source based on monolithic two-color laser diode," *Proc. SPIE 10684,Nonlinear Optics and its Applications* 2018,10684Y, 2018.

[117] B. Sumpf, J. Kabitzke, J. Fricke, P. Ressel, A. Müller, M. Maiwald, G. Tränkle, "785nm dual-wavelength Y-branch DBR RW diode laser with electrically adjustable wavelength distance between 0 nm and 2 nm," *Proc. SPIE 10123, Novel In-Plane Semiconductor Lasers XVI*, 101230T , 2017.

[118] N.Kim, S. Han, H.Ko,Y.A. Leem,H.Ryu,C.W. Lee, D. Lee, M.Y. Jeon, S. K. Noh, and K.H. Park, "Tunable continuous-wave terahertz generation/detection with compact 1.55 μm detuned dual-mode laser diode and InGaAs based photomixer," *Opt. Express*,19(16),15399,2011.

[119] M. Maiwald, C. Raab, W. Kaenders, B. Sumpf, G. Tränkle, "Monolithic dual-wavelength diode lasers with sub-MHz narrowband emission at 785 nm," *Proc. SPIE 10123, Novel In-Plane Semiconductor Lasers XVI*, 101230V , 2017.

[120] B. Breitenstein, M. Scheller, M. K. Shakfa, T. Kinder, T. Müller-Wirts, M. Koch, D. Selmar,Introducing terahertz technology into plant biology:A novel method to monitor changes in leaf water status, *Journal of Appl. Botany and Food Quality*,84,158-161,2011.

[121] R. Gente,M. Koch,Monitoring leaf water content with THz and sub-THz waves, *Plant methods:Biomedical central*,11:15,2015.

[122] D.Banerjee and W.Von Spege and M.D Thomson and S.Schabel and H.G Roskos, "Diagnosing water content in paper by terahertz radiation," *Optic Express*, 16,9060-9066, 2008.

[123] M.K.Shakfa,M.Scheller,B.Breitenstein,D.Selmar and M.Koch, "Monitoring the water status of economic plants with continuous wave terahertz radiation," *OSA Publishing*, 2009.

[124] N.Born and D.Behringer and S.Liepelt and S.Beyer and M.Schwerdtfeger and B.Ziegenhagen and M.Koch, "Monitoring plant drought stress response using terahertz time-domain spectroscopy," *Plantphysiol.* 164,1571-1577, 2014.

[125] E. Castro-Camus, M. Palomar, and A. A. Covarrubias, "Leaf water dynamics of Arabidopsis thaliana monitored in-vivo using terahertz time-domain spectroscopy," *Scientific Reports*,3,2930, 2013.

[126] M. Hamdi, J. Oden, J. Meilhan, F. Simoens, and B.Genty, "Enhanced Plant Water Status Measurement using THz Time-Domain Spectroscopy," *9th-thz-days*, 2017.

[127] A.Rehn,R.Gente,T.Probst, J.C Balzer and M.Koch, "Plant Water Status Monitoring with THz QTDS," *GeMIc*, 2016.

[128] I-M. Lee, N. Kim, E. S Lee, S-P. Han, K. Moon,and K. H. Park, "Frequency modulation based continuous-wave terahertz homodyne system," *Opt. Express*,23, 2,846-858, 2015.

[129] R.Wilk,F.Breitfeld, M. Mikulics, and M.Koch, "Continuous wave terahertz spectrometer as anoncontact thickness measuring device," *Applied Optics*,47, 16, 3024-3026, 2008.

[130] Carsten Brenner, Yinghui Hu, Jared Ombiro Gwaro, Nils Surkamp, Benjamin Döpke, Martin R. Hofmann , B. Kani, A. Stöhr, B. Sumpf, A. Klehr, J. Fricke ,"Near infrared diode laser THz systems," *Radio Sci.*,16, 2018.

Appendix A

Publications

Published publications and the conference proceedings from the start of the PhD research to the end.

Journal publications:

- Near Infrared Diode Laser THz Systems

  Carsten Brenner, Yinghui Hu, Jared Ombiro Gwaro, Nils Surkamp, Benjamin Döpke, Martin R. Hofmann , B. Kani, A. Stöhr, B. Sumpf, A. Klehr, J. Fricke - Adv. Radio Sci., Volume 16, 2018

- Compact continuous wave THz source based on monolithic two-color laser diode

  Jared O. Gwaro, Carsten Brenner, Bernd Sumpf, Andreas Klehr, Jörg Fricke and Martin R.Hofmann,Proc. SPIE 10684, Nonlinear Optics and its Applications 2018, 106841Y (14 May 2018).

- Terahertz frequency generation with monolithically integrated dual wavelength Distributed Bragg Reflector semiconductor laser diode.

  Jared O. Gwaro, Carsten Brenner, Bernd Sumpf, Andreas Klehr, Jörg Fricke and Martin R.Hofmann, IET optoelectronics special issue, Volume: 11, Issue: 2, 49 -52, 2017.

- Terahertz wave generation from dual wavelength monolithic integrated Distributed Bragg Reflector semiconductor laser diode.

  Jared O. Gwaro, Carsten Brenner, Bernd Sumpf, Andreas Klehr, Jörg Fricke and Martin R.Hofmann, ISBN 978-3-9812668-7-0. © IMATech e.V. Ratingen, Germany. GeMiC 2016, March 14–16, 2016, Bochum, Germany

Conference Proceedings:

- Compact continuous wave THz source based on monolithic two-color laser diode

  Jared Ombiro Gwaro, Carsten Brenner, Bernd Sumpf, Andreas Klehr, Jörg Fricke, Martin R. Hofmann - SPIE Photonics Europe, Strasbourg, France (2018)

- THz sources and systems based on near infrared diode lasers

  Yinghui Hu, Jared Ombiro Gwaro, Nils Surkamp, Benjamin Döpke, Carsten Brenner, B. Khani, A. Stöhr, B. Sumpf, A. Klehr, J. Fricke, Martin R. Hofmann - Kleinheubacher Tagung, Miltenberg, Germany, September 2017 (invited)

- Generation of Terahertz radiation with monolithically integrated dual mode Distributed Bragg Reflector semiconductor diode laser

  Jared Ombiro Gwaro, Carsten Brenner, Bernd Sumpf, Andreas Klehr, Jörg Fricke, Martin R. Hofmann - German Terahertz Conference, March 29-31,(2017).

- CW Terahertz generation with dual wavelength monolithic integrated Distributed Bragg Reflector semiconductor laser diode.

  Jared O. Gwaro, Carsten Brenner, Bernd Sumpf, Andreas Klehr, Jörg Fricke and Martin R.Hofmann, European Semiconductor Laser workshop, ESLW2016 ,September 23-24, 2016, Darmstadt, Germany.

- Terahertz frequency generation with monolithically integrated dual wavelength Distributed Bragg Reflector semiconductor laser diode.

  Jared O. Gwaro, Carsten Brenner, Bernd Sumpf, Andreas Klehr, Jörg Fricke and Martin R.Hofmann, Semiconductor Integrated Optoelectronics (SIOE) conference in Cardiff,UK 5-7th April (2016).

- Terahertz wave generation from dual wavelength monolithic integrated Distributed Bragg Reflector semiconductor laser diode.

  Jared O. Gwaro, Carsten Brenner, Bernd Sumpf, Andreas Klehr, Jörg Fricke and Martin R.Hofmann, Germany microwave conference, GeMiC 2016 ,March 14–16, 2016, Bochum, Germany

- Terahertz difference frequency generation by a monolithic integrated dual mode Distributed Bragg Reflector semiconductor diode laser.

  Jared O. Gwaro, Carsten Brenner, Bernd Sumpf, Andreas Klehr, Jörg Fricke and Martin R.Hofmann, 7th International Workshop on Terahertz Technology and Applications Workshop, Kaiserslautern 15-16th March, (2016).

www.ingramcontent.com/pod-product-compliance
Lightning Source LLC
Chambersburg PA
CBHW070738220326
41598CB00024BA/3471